THE BODY BUILDERS

THE BODY BUILDERS

INSIDE THE SCIENCE OF THE ENGINEERED HUMAN

ADAM PIORE

ecco
An Imprint of HarperCollins*Publishers*

The author first detailed portions of Hugh Herr's experience (chapter 1), the Patrick Arnold story (chapter 2), Stephen Badylak's research on "pixie dust" (chapter 3), and the story of the ARO project overseen by Elmar Schmeisser (chapter 6) in articles originally published in *Discover* magazine. In describing the work of Gordana Vunjak-Novakovic (chapter 3), the author drew from material originally published in *Columbia Magazine*. In describing Phil Kennedy's self-experimentation (chapter 6) and the work of Emad Eskandar and Darin Dougherty (chapter 8), the author drew upon work he had previously published in *MIT Technology Review*. Derek Amato's story (chapter 9) originally appeared in an article written for *Popular Science*.

FIRST EDITION

Designed by Joy O'Meara

Library of Congress Cataloging-in-Publication Data has been applied for.

ISBN 978-0-06-234714-5

17 18 19 20 21 LSC 10 9 8 7 6 5 4 3 2 1

To my grandparents:
Mannie and Nora Piore, who asked the questions,
and Polly and Sid Kline, who told the stories

CONTENTS

INTRODUCTION 1

PART I: MOVING

CHAPTER 1 THE BIONIC MAN WHO BUILDS BIONIC PEOPLE
Replicating the Way We Move 13

CHAPTER 2 THE BIRTH OF BAMM-BAMM
Decoding the Genome and Rewriting It 53

CHAPTER 3 THE MAN WITH THE PIXIE DUST
Regenerative Medicine and the Quest to Regrow Limbs 95

PART II: SENSING

CHAPTER 4 THE WOMAN WHO CAN SEE WITH HER EARS
Neuroplasticity and Learning Pills 137

CHAPTER 5 SOLDIERS WITH SPIDEY SENSE
Intuition and Implicit Learning 179

CHAPTER 6 THE TELEPATHY TECHNICIAN
Decoding the Brain and Imagined Speech 219

PART III: THINKING

CHAPTER 7 THE BOY WHO REMEMBERS EVERYTHING
Viagra for the Brain 253

CHAPTER 8 THE SURGEON CONDUCTING THE SYMPHONY
DBS and the Power of Electricity 289

CHAPTER 9 SUDDEN SAVANTS
Unleashing the Inner Muse 313

CONCLUSION 339

Acknowledgments *349*
Notes *353*
Index *367*

INTRODUCTION

This is a book about science and medicine. But the journey that led to it began about as far away as one can get from the antiseptic, data-driven confines one normally associates with these pursuits. In fact, it began so far from a laboratory or operating room that I'm almost reluctant to chronicle it here, for fear I will drive away some of the readers hoping to delve into the mysteries of disciplines such as neuroscience, biomechanics, genetic engineering, and the like. Those readers will have to trust me. We'll get there.

For me, this book began on a sunny hillside atop the campus of the University of California, Santa Cruz, in the 1990s. I was sitting cross-legged, surrounded by other students, gazing out across green athletic fields and down over redwood-studded hills upon the blue, pristine stillness of the Pacific Ocean. The curving, rocky coastline of Monterey Bay stretched out before me. It was the kind of view that calms the soul and evokes endless possibilities and adventure.

I was a freshman in college and officially "in class." But this was like no class I had ever attended before. I thought of my friends back east, sitting inside, soon to be snowbound. Of high school malaise, feeling penned in and hearing the Charlie Brown drone of a teacher as I daydreamed of other places. Places like this. It seemed amazing to me that I should be here, "in class." Who knew this was possible?

But then, possibilities were the point. The class was called "Humanistic Psychology," and my TA, a man named Jim Brown, thought this was an appropriate setting to introduce us to its optimistic, unapologetically new-agey focus. It centered on something called "the human potential movement," a kind of psychology nurtured on the stoned utopianism and creative anarchy of the 1960s counterculture. I was eighteen, still carrying the remnants of my own special blend of adolescent miseries, resentments, and fears. And I was spellbound.

Humanistic psychology was about human transformation, about breaking free of that which limits us and holds us back. It was a discipline that was in part a reaction to the often pessimistic approaches of traditional psychoanalysis and behaviorism, approaches that focused on understanding what caused the neuroses that warp the way we perceive the world or cause us to behave pathologically. Humanistic psychologists like Abraham Maslow were interested in the next step. What did it look like when individuals were able to let go of their disappointments, their fears, and transcend the wounds of the past? All of us are motivated, if given the choice, Maslow argued, to achieve our full potential as human beings—to "self-actualize," to find happiness, embrace creativity, find nurturing relationships, and move beyond the things that hold us back. To study this, Maslow looked not to those among us who were suffering, but to those who were thriving. What did they have in common? And how was it that they had become "self-actualized"?

"It is as if Freud supplied us the sick half of psychology and we must now fill it out with the healthy half," Maslow wrote in 1968.

Later, when I became a journalist and foreign correspondent, and first interviewed survivors of Pol Pot's genocide in Cambodia, I began to question the value of Maslow. One afternoon, standing on a rutted road in the countryside, I asked a stooped, toothless beggar what her thoughts were on a United Nations tribunal to try the surviving leaders of the murderous Khmer Rouge regime. Tears began to stream down her face. "They killed my children," she said. "And that is why I am like this."

What did Abraham Maslow have to teach someone who had been through something like that, I wondered? How was "self-actualization" even relevant—not just for her, but for everyone else—in a place where such injustices could occur? There was nothing about what I had studied in college, it seemed then, that applied in a country like this.

But eventually I would come to think about these questions in a new way. I had arrived just as Cambodia seemed to be emerging from thirty years of civil war. The people I met were still grappling with the legacy of the brutal four-year period in the 1970s when one in four individuals died of starvation, murder, and disease—a cataclysm that had ripped apart the entire society and left all who survived traumatized and broken. The stories I heard were so tragic they often brought me to tears.

Yet the traditional Khmer New Year that first spring in 1999 was a huge, festive outpouring. Just the year before, people had been afraid to venture out, in the wake of the 1997 coup d'état. Now they thronged the streets and alleyways outside my apartment in downtown Phnom Penh, dancing and eating and celebrating. I had seen the pictures of hollow-eyed refugees. Now I watched children roam free. The local mayor had transformed a washed-out patch of dirt next to the Mekong River from a muddy morass into a park with lush grass and flowers, a place where families could picnic. The survivors around me were not grim; they were joyous. You could see it in their faces, euphoria and relief, renewal and resilience. And I found its appearance in what I had mistakenly taken to be a hopeless desert of sadness and loss, far more powerful than anything I could have imagined.

How was this joy possible? Even the most unimaginable calamity, it seemed, had not crushed the humanity of the people around me, and the ability to embrace the moment and each other. In fact, it seemed to make it burn even more brightly. Where did this remarkable resilience, this joy, come from? And why did it speak so powerfully to me?

Sometimes the most inspiring stories, the ones that show us what we're really capable of, the ones that show us what is really important, emerge from the worst tragedies we can imagine. It wasn't just "human potential" that had so fascinated me on that hillside in California, I realized, but the resilience of the human spirit. That inherent urge, instinct, momentum to find, when we lose a part of ourselves, a way to be whole again—not just to survive but to thrive. I've always been fascinated by what it is that makes people who have experienced setbacks able to move on and look ahead.

You may ask what any of this has to do with those long words I alluded to in the beginning of this introduction. How could neuroscience, biomechanics, and genetic engineering possibly relate to Pol Pot or to humanistic psychology?

I have chosen to write a science book, not a story about psychological trauma, Cambodian history, or Abraham Maslow's hierarchy of needs, because today some of the most extreme and exciting examples of the triumph of human potential and resilience are being unleashed by medicine and science. This is something I experienced quite by accident. After returning home from my Cambodian adventures, working at *Newsweek,* covering 9/11, traveling to Iraq, I met a bionic engineer, named Hugh Herr, with a story that so intrigued and inspired me, it made me look at science and technology in a new way. I followed the thread of the story he was part of—one involving neuroscience, biology, and amazing advances that are transforming what is possible. And as I got to know more people whose lives had been profoundly affected by the quiet revolution occurring in labs and clinics around the nation, their stories began to work the same kind of ineffable magic I first encountered on the streets of Phnom Penh in the 1990s.

This book is largely about a field called "bioengineering," and the way that scientists, doctors, and sometimes patients themselves are using it to unlock resilience in human bodies and minds that previous generations could only have guessed was there. Though the subjects I explore highlight what could turn out to be some of our era's most dramatic scientific feats in neuroscience, regenerative medicine, pharmacology, and bionics, this is not meant to be a clinical, intellectual tome about the workings of the human body and mind. This is a book about people who refuse to give up. When I set out to report it, I went looking for people helping themselves and others regain things they thought were gone forever—the ability to run and dance, to see a mountain in the distance, to recognize a loved one, even simply to communicate. The things, in other words, that makes us feel most human.

Last century, we reached a tipping point of large-scale engineering—an explosion of physical ingenuity that resulted in stunning feats of mechanical and structural prowess, and triumphs over the normal large-scale limitations imposed by the physical world: the construction of the Empire State Building, the invention of the airplane, the lunar landing. Today engineers are training their sights inward. The new frontier is the human body, and the insights that scientists and these modern-day builders and architects are realizing are helping us restore lost function to the injured and unlock new potential in the rest of us.

It's a newsworthy topic. In recent years, "technology," as one introductory biomedical engineering textbook puts it, "has struck medicine like a thunderbolt." That technology is allowing scientists to unleash untapped powers in the human body that we are just beginning to understand. Stem cells that can rebuild lost body parts. Brains that can rewire, detouring around the site of a catastrophic injury. Ideas and perceptions just outside awareness that contain the sum of the wisdom of all of our experiences.

Some of the technologies I will detail sound like something out of a science fiction movie. While reporting this book, I visited people who have regrown fingertips, and leg muscles that were blown off by explosions. I met a woman who can "see" with her ears and people trying to give "locked in" patients who can no longer speak the ability to communicate telepathically.

But the same transformative technologies that are making all of this possible also raise difficult questions. New technologies are allowing scientists to reverse-engineer the human body and mind with unprecedented precision, to take apart and analyze its disparate parts and figure out how they work together down to the level of molecules and back up again, to build replacement parts for those who have lost them. But why stop there? Many scientists in fact are actively exploring how we might use these same technologies to help the able-bodied exceed their natural-born limitations. If we can fix the human body and mind when it breaks, why not build better versions of ourselves? Why not enhance, augment, transcend? Why not see how far we can push it?

As the author of a recent report to the European Parliament noted, the technologies involved in human enhancement, such as genetic engineering, bionics, and drugs that boost brain power "signal the blurring of boundaries between restorative therapy and interventions that aim to bring about improvements extending beyond such therapy."

"As most of them stem from the medical realm, they can boost societal tendencies of medicalization when increasingly used to treat non-pathological conditions," the report noted. In other words, the function of medicine in our society might shift and expand—possibly in fundamental ways. This, some warn, could have a wide array of unintended consequences, ranging from widening inequality between the rich and poor, to an arms race of neurological and physical enhancements. Some warn it may even alter what it means to be human altogether and cast into doubt the very basis of liberal democracy—the belief that all human beings are equal by nature.

"The original purpose of medicine is to heal the sick, not turn healthy people into gods," Francis Fukuyama writes in his book *Our Posthuman Future: Consequences of the Biotechnology Revolution*. Yet the allure is hard to resist. We human beings have been attempting to transcend our natural limitations since the very dawn of civilization itself. Ancient Greek Olympians are said to have chewed raw ram testicles to bulk up, long before anyone could have told you that the testicles are a great source of testosterone, the male hormone that promotes muscle growth, bone mass, and strength. Writers and scientists have been using caffeine and nicotine to promote mental focus since at least 600 B.C.—even though, back then, nobody knew that caffeine blocks a brain chemical called "adenosine," which promotes sleep and suppresses arousal. Or that nicotine mimics the neurotransmitter acetylcholine, which causes brain cells in our sensory cortex and areas involved in attention to snap to life.

In others words, even without a basic understanding of biology, physics, and chemistry, humanity for millennia has worked relentlessly to hack nature and manipulate body and mind. Our resulting technologies have often been crude, mysterious, and unreliable. What happens now that this is changing? Should we be worried? Is this a good thing?

It's a question that a number of medical ethicists have highlighted as one of the most pressing of our time. But it is certainly not a new concern. Those same Greeks who gave us the testicle-chewing Olympians also left behind the myth of Daedalus, who used his engineering prowess to build wings of wax and taught his son Icarus to fly, defying the gods and bringing calamity down upon his house. Will we as a society, drunk on our own ingenuity, fly too close to the sun? Will we use enhancement as a tool of repression or war?

I must admit that while I researched this book, those concerns were often overshadowed in my own mind by the sheer wonder I experienced when I actually encountered some of the technologies and considered what it would be like to actually be "augmented."

Like when I put on a "muscle suit" that allowed me to lift weights with my fingertips. It felt like I was picking up a piece of paper. Or when I spoke with a boy with a memory so sharp that when he was two, he could recite the numbers on the inspection stickers of every car in his neighborhood. And then spent the day with a man working on a pill that will give us all that kind of memory.

I was repeatedly seduced, like many of the scientists themselves, by the urge to see how far we might take it. But the specter of excess, that our own technologies might somehow undo us, always rose up again. When I told people about my adventures, they often asked me, "Is this something we should embrace?" I wondered if I should even try to answer that in this book. After all, "it depends." As one military scientist sagely shrugged when I asked for his thoughts: "Is a baseball bat a good thing or a bad thing? It's a good thing if you use it to play baseball, but not if you use it to beat somebody over the head." In my own way, I have done my best to provide some answers. After reading this book, you'll at least be far better equipped to come up with your own answers. You'll know more about what might be possible and why, and you'll understand the debates when you encounter them.

At its heart, however, this book is not about ethical questions, technical specifications, or even the scientific discoveries that make the stories I'm about to share newsworthy. This is a book about people. Many of them started out like you and me but found themselves, through bad luck, their own choices, or the inevitable pitfalls of life, facing challenges many of us can hardly imagine. In my mind, their resilience makes the pursuit of new technologies a worthwhile undertaking, if not a necessary and noble one.

The characters in this book—the scientists and those they are trying to help—show all of us something about ourselves. Where our limits lie. What we are capable of, and how. In that sense this book is not just about human resilience. It must, inevitably, also be about the possibility of transcendence.

The book is divided into three sections (and multiple chapters within each) organized around efforts to understand and reengineer human movement, sensing, and thinking.

But always people and their stories will center us. And they are amazing stories. So let's start with one of the most remarkable people I met: Hugh Herr.

PART I

MOVING

CHAPTER 1

THE BIONIC MAN WHO BUILDS BIONIC PEOPLE

REPLICATING THE WAY WE MOVE

It had already begun to snow when Hugh Herr and Jeff Batzer set off up a wooded trail near the base of New Hampshire's Mount Washington on a frigid morning in January 1982.

They'd been planning their trip for months and had driven through the night from Lancaster, Pennsylvania, to get there. Herr, a baby-faced seventeen-year-old with a floppy mane of brown hair, knew Batzer was eager to get to the top. But as the two climbers hiked in, they were still unsure whether they would attempt to summit that day. The crown of the mountain had been obscured by ominous clouds when they'd left that morning, and after a twenty-five-minute hike through a deep, narrow gorge it was blocked from their view.

The pair halted about three-quarters of a mile in at the base of Odell's Gully, a notorious ice field where just months before a young climber had tumbled to his death. As Herr and Batzer stood gaping up at a long, blue ice runnel snaking down steeply from a broad elevated plateau, visibility remained good, and the frigid wind whistled faintly around them. They shrugged off their packs, dumping their bivvy gear at the side of the trail so they could go up fast and light, unencumbered.

Herr led the way up the steep ice wall. At seventeen, he was three years younger than Batzer. But there was never any question he'd go first. Herr had been climbing with his older brothers since the age of seven. By the time he'd reached his teens, Herr was a nationally recognized rock climber, a "child prodigy" to climbing peers, considered one of the nation's top ten, and possibly the best on the East Coast.

Just months before, in fact, Herr had pulled off a climb so audacious, so technically challenging, many in the climbing community had at first refused to believe the news. Herr had taken aim at Super Crack, considered perhaps the hardest climb in the entire Northeast. The climb consisted of an angled pinnacle, split by a narrow, inch-and-a-half-wide crack running from bottom to top, angling outward as it went up. Halfway up, an intimidating eighteen-inch overhang completely blocked the route, requiring climbers to hang one-handed and through a gravity-defying miracle somehow reach around the overhang to grip an impossibly spaced hold far above it. The first climber ever to successfully make the ascent, in 1972, fell thirty-two times before he reached the top. The year before Herr's attempt, one of the world's top climbers, Kim Carrigan, took an entire day on the wall to conquer it. The news had electrified the climbing community, nonetheless—Carrigan was the first climber ever to accomplish this feat in such a short period of time. Most climbers had to "siege" for several days.

To prepare for his climb, Herr carefully studied the contours of the wall, the faint ridges, and sparse handholds. Then he built a full-scale replica in his barn out of cement block, wood, and mortar and trained for an entire winter, attempting to climb his replica wall several times a day. By the time spring rolled around and Herr attacked the real Super Crack, the route was so familiar, Herr so meticulously prepared, he almost made it up on his first try, before falling. Then he started up again at the bottom and completed the climb in less than twenty minutes. It wasn't for nothing that some called Herr "the Boy Wonder."*

Now on that frigid day in January 1982, just a few months later, Herr was hammering into the sheer ice face of Mount Washington with picks and crampons. He tied himself into an ice screw placement, trailing ropes, and began to belay Batzer below him. As he climbed the steep face, Herr eyed the huge walls of snowpack above them cautiously, aware of avalanche danger, and hugged the edge of the gully to one side.

The climbers reached the top of the gully at about 10 A.M, just as the weather conditions had begun to shift. The wind howled furiously around them, so loud they had to duck behind a boulder to hear each other as they conferred. They were just 1,100 feet below the summit, about a mile hike across far easier terrain than what they had just conquered.

"Do you want to try for the summit?" Herr asked.

"Think we could make it?"

Stepping out from behind the boulder and back into the howling winds, they started up again, hoping to bear the storm. They moved at a slow trot, crouching into the wind. But the temperature had fallen to just above zero, and the wind was soon gusting up to 94 miles an hour, deafening and violent, slapping them with frigid, stinging fingers. Visibility had dropped to five feet. Batzer would later recall the snow coming at him almost horizontally, and having a terrifying sense that if he jumped, the fierce winds would simply

---

* For more on Herr's early climbing career see Osius, *Second Ascent*.

pick him up and hurl him fifteen feet through the air. It was too much. Within just a few hundred feet, Herr and Batzer had to shout over the noise to be heard. "Let's get out of here!"

As the two turned to go back, they were in whiteout conditions. Herr could barely see his own two hands. The terrain was somewhat flat, at most gently rolling, and thus each direction was the same as the next given the lack of visibility. All they could do was calculate their route toward the warm safety of civilization based on the direction from which the winds had been nipping at them all the way up. What they didn't know was that the wind had shifted. Instead of heading back the way they had come, the climbers unknowingly began to descend into a gully system that looked deceptively similar to their initial descent plan.

"It was," Herr would later recall, "a white maze from hell."

By the time they realized they were in the wrong place and stopped to confer, it was too late to turn back. The winds above them, they both agreed, had grown so ferocious they no longer seemed survivable. So the two boys continued their descent, hoping for the best.

Though they didn't know it, Herr and Batzer were on the edge of a vast wilderness area. And they were hiking straight into its teeth.

Yet it seemed so peaceful at first. Down in the trees, the wind faded, precious calm returned, and gentle flurries fell silently around them. Soon, however, Hugh Herr and Jeff Batzer found themselves struggling through snowdrifts up to their chests, picking their way across unfamiliar iced-over streams, past boulders strewn like toy blocks amid towering fir trees. As the daylight faded, they hiked on, picking out a path next to a stream that seemed to be growing wider. They didn't have much choice—the snow was so deep in most places, it reached the tree limbs, which meant the boys would have had to tunnel under it to avoid walking into the branches if they strayed too far from the riverbanks. But the path also added an unanticipated element of danger: Twice that first night, Herr stepped through the ice and felt the frigid water surge up to his knees, soaking his hiking boats and dunking his lower limbs in bitterly freezing depths.

Still the pair hiked on. They hiked for hours into the night to try to keep warm, until eventually they collapsed beneath a boulder, hugging each other for heat, covered with branches plucked from pine trees. The boys carefully removed their boots. Batzer shared some of his layers with Herr, whose own falls had left everything below his waist waterlogged and frozen.

The next morning the boys set out again at first light and hiked all through that second day, stumbling onward, hopelessly trapped, their desperation growing along with their exhaustion. By midday their legs seized in painful cramps. The river water that remained in Herr's boots froze his socks into hard chunks of ice, while the perspiration in Batzer's shoes also hardened. All that ice accelerated their descent into hypothermia and frostbite, while the deep snow hindered their movement.

By day three, the boys were badly dehydrated, and weak. Herr's feet had grown so numb, he could barely balance, and, Batzer recalls noticing with alarm, he had grown ominously quiet. The two friends crawled under another rock and tried to get warm. Batzer struck out on his own in a last-ditch effort to find help, but made it less than one mile before turning back. By the end of the day, the boys had begun to accept the stark fact that they probably weren't going to make it out alive. Years later, Batzer would recall asking Herr about his faith and if, at seventeen, he was ready to die. Both made peace with their maker.

It would be three days before a woman in snowshoes would stumble on their tracks. She found the boys huddled and frozen still beneath that boulder, hours from death. By then an avalanche had claimed the life of a search-and-rescuer out looking for them, and both Herr and Batzer had severe frostbite.

By the time the boys arrived at the hospital, Batzer's body temperature had fallen to 90 and Herr's hovered around 91. Doctors amputated four of Batzer's fingers, a thumb, part of his left leg, and all the toes on his right foot. Herr wasn't as lucky. The doctors amputated both his legs below the knee. The promising athlete, the rock-climbing legend in the making, despite his iron will, despite his fearlessness, would never be whole in the same way again.

Back home in Pennsylvania, in those devastating early days after the doctors had amputated his legs, Hugh Herr had a recurring dream.

He was running through the cornfields behind his parents' house, going impossibly fast, the sun and the wind on his face; he was almost flying. The ineffable sensation of freedom would remain vivid decades later. Then he would wake up to the stumps of his legs below his sheets. He'd feel that lump of loss rise up in his throat, a hollowness in his chest. And he'd remember the whiteness and howling winds with a shudder. The doctors had told him he would never run or climb again.

Herr's first prosthetic legs were made of plaster of Paris. They were stiff and lifeless weights on the ends of his stumps. The prosthetist who fitted them suggested Herr might one day be able to walk without canes, but that was about it. He could drive a car using hand controls. But his rock climbing days were over.

Herr was deeply depressed. But he wasn't beaten. And soon he had cabin fever. One morning Herr rolled out of bed and pulled himself around the room by his arms, seeing what he could do. Not long after that, Herr pushed himself into the empty kitchen on his butt. Hefting himself up onto a chair, Herr climbed onto the counter, reached up to grip the top of the family refrigerator, and, hanging by both hands, swung his torso and stumps from one side of the refrigerator to the other, as if dangling from the Super Crack overhang. Then he did ten pull-ups.

Herr lowered himself to the kitchen floor and scooted over to the basement door, pulling himself down the stairs, then climbing back up their underside. After reaching the top and climbing back down again, he collapsed in a heap on the cold cement floor, laughing in relief. It was the first time he had been in a fully upright position since the accident, Herr would later tell his biographer, Alison Osius, a member of the U.S. Climbing Team and longtime writer and editor at *Climbing* magazine. Herr's legs were gone. But no one was going to tell him he couldn't climb. No one, no way.

Seven weeks after doctors amputated his lower legs, Herr climbed into a car with his older brother, Tony, and headed to a series of cliffs along the Susquehanna River. For years he had been doing things on rock faces that people said he couldn't, but even he was amazed by his performance that day. Weak and recovering from the surgeries, Herr was wobbly on his new artificial feet. On the rock wall, even with the artificial limbs on, it was a different story. "I felt more natural scrambling on all fours than walking," he says.

By summer Herr was in a local machine shop experimenting with his artificial limbs, tricking them out for the climbing wall. Every few weeks he headed to Philadelphia to meet with prosthetist Frank Malone for refitting and adjustments. Herr had begun tinkering with the design of his new legs, adjusting the length and playing with different materials to make them lighter.

"I realized that my prostheses need not look human," he says. "They are a blank slate: I could create any prosthetic device I wanted for form, function, and enhancement."

Herr felt silly putting climbing shoes on the ends of his prostheses. So he threw the shoes out and glued climbing rubber directly to the bottom of his mechanical feet. Then he went to work on their shape. For expert pitches where he planned to stand on small rock edges the width of a dime, normal feet were a disadvantage. So he designed a prosthesis about the size of a baby's foot. He created a pair of feet with toes made of laminated blades that he could jam into tiny rock fissures far too slim to hold a normal human foot. He created another pair of spiked feet that allowed him to climb ice walls as if they were rock. He made the height of the legs adjustable; at 7 feet, 5 inches, he could reach handholds and footholds far beyond the range of any able-bodied climber. Herr drilled holes throughout legs fashioned from aluminum tubes, making them so light they just barely supported his weight, while increasing the number of pull-ups he could do with them on and the distance and speed with which he could climb.

"Through technological innovation, I returned to my sport stronger and better," he says.

The other climbers began to gather in small groups at the base of the craggy, conglomerate cliffs just to watch Hugh Herr work. A hundred feet above the Susquehanna, he'd be moving across the sheer rock face, his toned biceps and shoulders running with sweat in the midafternoon sun, his face a mask of concentration, his biological legs ending in stumps a few inches below his knees, transforming into the bizarre contraption of the day, metal glinting in the light. Herr's body was lighter. He was faster. He could climb at angles and on spaces no one else could. Hugh Herr wasn't disabled. He was augmented.

It was only a matter of time before Herr began asking an obvious question. If he could, just with a little tinkering, transform his legs on the climbing wall, what might he accomplish if he set his mind to upgrading the legs he relied on in the horizontal world? If he could build legs that allowed him to reach rock handholds normally out of range, what else might he create?

It's a cold, drizzly day as I make my way across the redbrick walkway of Kendall Square in Cambridge, Massachusetts, and head to Hugh Herr's office in a sleek modernist building at the Massachusetts Institute of Technology (MIT). It's been more than twenty-five years since his accident, and since the boy wonder impressed the other climbers with his modified aluminum legs. And there will be no alpine climbing moves for me to see today. Instead, soon after I arrive, Herr performs a physical feat that is, in fact, far more memorable: He *stands up* from his chair, and puts on his jacket. Then he leads me first down the stairs and, in a brisk, determined gait, across a wide open, snow-covered quad.

Herr is wearing expensive Italian shoes and a green puffy jacket, his "legs" obscured by his designer jeans. I'm trying to avoid ice, and tripping over uneven ground, and he's chatting casually about restaurant options. If I didn't know about the accident, if I couldn't hear that faint metallic squeak that seems to speed up and slow down with every change in Herr's determined pace, it seems to me it's possible I might never know that this trim, athletic man with the full head of hair and obvious vitality would in another age—in any other age—be considered profoundly disabled.

But then, one of the first things Herr did when I arrived was fix me with an inscrutable poker face and hike his neatly ironed pant leg up to show me just how far he has taken his engineering adventures since the early days tinkering with his prostheses all those years ago.

Five inches below Herr's knees, at the point where the doctors amputated, Herr's natural legs transform into aluminum an inch in diameter, atop sleek masses of silver gears and wires, which power flat black feet resembling the bottoms of flip-flops.

Each bionic limb contains three internal microprocessors and a quarter-size inertial measurement unit (originally designed for missile guidance systems) that track and adjust the foot's location in space and react to changing terrain and different walking speeds, allowing Herr to push off the ground with seven times as much power as is possible with the best of its predecessors, all while expending less energy. Herr's legs are motorized and capable of adjusting 500 times a second for angle, stiffness, and torque. Herr calls them "wearable robots."

"Someday soon, these kinds of devices will be no more unusual than a pair of glasses many of us now use to improve our vision," he tells me.

It's a stretch, of course, to compare Herr's bionic limbs to a pair of glasses. Because if the hardware sounds impressive, complicated, and high tech, the software—and beyond that, the actual data used to determine the precise movements of the many constituent parts that make up these cutting-edge robots—is even more so. Herr's initial prosthetic creations hammered together all those years back for the climbing wall deviated from nature's designs in myriad, random ways: extensions that grew his limbs to 7 feet, 5 inches, baby-sized feet, bladed toes. Herr embraced liberation from the human form, and the option of experimentation. He dreamed of someday wearing wings on his lower limbs. But Herr's pursuits since arriving in Cambridge to pursue an MIT degree twenty years ago have unexpectedly led him in the opposite direction.

By deviating from biological precedent as he explored the engineering challenges posed by movement, Herr came to appreciate the sublime complexity and genius of the natural human form. And Herr realized that the key to augmenting the body lay first in reverse-engineering and deciphering nature's solutions on the most minute level. Only then did it make sense to begin to build on them. In the process, he realized, he could help a lot of people.

Today Hugh Herr is one of the world's leading prosthetic designers, creating devices that both restore function to the disabled and augment the abilities of the able-bodied. What Herr and others are learning about human movement—the precise way that our ligaments, tendons, and muscles store, transfer, and release energy—is transforming the way we think about our own innate limitations.

And this has allowed Herr to do something he'd long since accepted as impossible.

Hugh Herr is walking—*really walking*—again.

One way to look at the human body and the constituent parts we use to move is as a simple pulley system, perhaps the kind used to move a marionette. The bones are the scaffolding that gives the body shape. The muscles and tendons are the pulleys that move the scaffolding, jerking them this way and that. The ligaments are what hold it all together. That's how Herr climbed those mountains, how I lift and carry my four-year-old daughter and then place her gently in her bed when she falls asleep, how all of us can glance around us, take in our surroundings, and then reach out to act upon the world. Pulleys.

But look a little closer and what might seem at first to be a simple system reveals itself to be far more complicated. These basic components, joined in large numbers, form an intricate web, capable not just of directing movement in real time, but also of storing, transferring, unleashing, and then recovering the unseen ingredient that makes all movement possible—energy. The way the body's myriad joints, tendons, muscles, and bones move in concert to juggle, store, and unleash this elemental property has been honed over millennia for maximum efficiency. And for centuries, the exact way that these disparate parts fit together and combine to generate, store, and release the energy we need to heave a rock, to sprint across the plains, or even to conduct that most basic of human abilities—bipedal walking—has been obscured from our view by its confounding complexity. If the human body is a pulley system, it's one made of rubber bands—and those elastic bands are knitted together far more intricately than the relatively symmetrical geometry of a spider's web.

Simply to replicate the precise movements of a human arm in two dimensions requires simultaneous measurements from twenty-nine different muscles that work together to counterbalance one another, notes Patrick van der Smagt, who directs the biomimetic robotics and machine learning lab at Germany's Technical University of Munich and was part of a team that attempted to build a bionic arm.

There are, according to most estimates, somewhere in the neighborhood of 206 bones, 360 joints, 700 muscles, 4,000 tendons, and 900 ligaments in the human body. For much of human history, no one had the tools to measure them—and certainly no one had the tools to emulate them—effectively. They were the stuff of art and sculpting more than science; there was no medical equivalent to Leonardo da Vinci's Vitruvian Man, which captured the proportions of man in a circle.

Hugh Herr was confronted with this reality in the most visceral and devastating way after his accident. Before he'd lost his lower limbs, he hadn't thought much about the way the human leg bounces, or explodes off the ground to generate movement. But in those early days after the accident when Herr tried to adjust to his new plaster of Paris "legs," it was hard to think about anything else. They were stiff, lifeless weights on the end of his stumps that dragged him down rather than propelled him up, the way normal legs do. Herr was lucky at first: He could escape to the climbing wall, a refuge from the normal laws of terrestrial gravity, where he could flourish. Thousands of feet up, he could look out across acres of craggy rock faces, across panoramic vistas of verdant valleys and streams, and he felt as free as he'd ever been.

But eventually even that wasn't enough. Herr's rigid prostheses lacked the natural cushioning provided by the tendons of the ankles and feet. The sockets where his legs met the artificial limbs chafed when he walked, leaving him raw and bloody and putting powerful strain on his knees. Some of the legs had fake feet that fit in shoes, and the plastic was sometimes skin-colored and made to look like the real thing. But, the truth was, these rodlike prostheses weren't much better than a peg leg you might see a Civil War veteran wearing in the nineteenth century.

The more Herr rediscovered his freedom on the vertical wall, the more frustrated he grew with his limitations on regular ground, and the more convinced he became that he had to do *something.*

"The medical community was giving me these devices and saying, 'This is the best, live with it,'" Herr says. "I just couldn't accept that what they gave me was really the best that we could produce."

One day, instead of just trying to make a better climbing leg, Herr began trying to make a leg that wouldn't hurt so damn much when he walked. First he tried stuffing the sockets with leather and rubber to cushion them. Then Jeff Batzer, Herr's fellow Mount Washington survivor, introduced him to a prosthetist and orthotist named Barry Gosthnian, who agreed to try to help him take his ideas further. Together the two brainstormed. Gosthnian had been an air force mechanic in Vietnam and recalled the shock-absorbing hydraulic supports used in aircraft landing gear. Perhaps, he suggested, a hydraulic cushion of some sort could soften the impact in the socket.

That fall, Herr enrolled at Millersville University, a state school in central Pennsylvania. He'd always been an average student, given to Cs and even the occasional D—uninspired by the classroom. But now Herr had a reason to attack his studies with the same focus and intensity he'd harnessed to conquer Super Crack. The innate precision and patient fortitude Herr had displayed when constructing that giant replica wall in his parent's barn, he now applied to math and physics.

By the time he graduated, Herr shared a patent with Gosthnian for a cushioned socket with inflatable bladders to cut down on the painful chafing caused by his prostheses. The bladders, made out of soft, flexible polyurethane membranes, were located wherever weight-bearing portions of the leg stump pressed against the socket, cushioning the force and softening the pressure on the stump as needed.

Herr, meanwhile, had also completed a climb from the depths of academic apathy to heights every bit as impressive and surprising as his famous climb up Super Crack. The former C and D student, who had for years whiled away classroom time dreaming of climbing, was now a straight-A student. In addition to good grades and a patent, Herr also had an acceptance letter to a graduate program at MIT in mechanical engineering.

In the fourth century A.D., the Roman military expert Publius Flavius Vegetius Renatus wrote in great detail about one of the most fearsome machines of his day: the catapult. The uses for this versatile weapon and its projectiles were many. You could knock down legions of opposing soldiers like bowling pins. You could batter in city walls. The Mongols liked to hurl the infectious corpses of bubonic plague victims into town squares to terrorize the local citizenry.

The device itself was an engineering marvel. Yet one of its key components had been around for as long as mammalian life itself: Vegetius noted that the very best springs, the oversize rubber bands the machines used to hurl deadly projectiles through the air, were fashioned out of tendons pulled from the necks of oxen. Others used braided ropes made from Achilles' tendons.

Modern-day scientists sometimes look back with wonder on this early wisdom and smile ironically—several mentioned it to me while I reported this book. Despite there being widespread knowledge of their uncommon properties, it would take sixteen more centuries before we would even begin to understand the crucial role the uncommon elasticity of tendons plays in the biomechanics of human and animal locomotion.

Perhaps appropriately, it was another Italian who made the discovery that started it all. In the 1950s, Giovanni Cavagna, a physiologist at the University of Milan, recruited human subjects to run on a treadmill equipped with weight-sensitive plates capable of recording the amount of force exerted on the ground with each stride. By estimating their carbon dioxide output and oxygen consumption (which was possible thanks to previous experiments that measured them using a mask), Cavagna was able to calculate the number of calories they were burning with each stride, and compare that to the amount of force each jogger had produced. Cavagna was surprised by the results. His subjects weren't consuming nearly enough oxygen to account for all the energy they appeared to be generating with each stride. If his calculations were correct, and Cavagna surely double- and triple-checked them, that meant the extra stride energy had to be coming from somewhere else. Cavagna offered a revolutionary theory: Almost half this power, he suggested, came from "elastic recoil energy" within the leg—some form of dynamically stored oomph that could provide an extra bounce. Somehow, he suggested, the leg was behaving like a spring.

It was just theory until, soon after, the British zoologist R. McNeill Alexander stumbled quite by accident on the first clues suggesting how these springs worked. Alexander had noticed that horses occasionally broke their legs jumping over hurdles, and he wondered why. How much strain, he wanted to know, did the act of jumping actually place on the lower limbs of mammals? And how close did that strain bring the leg to its breaking point?

To find out, Alexander taught a German shepherd named Happy to bound down the long hallway outside his lab and leap up onto an elevated platform. Just before Happy jumped, floor plates recorded the force exerted by his limbs on the floorboards while a camera recorded the position of all the constituent parts of Happy's hind limbs (marked with reflective tape and a felt marker). By feeding all of this data into known mathematical equations, Alexander was able to calculate the amount of force exerted on each disparate part of Happy's leg as he propelled himself through the air.

With this data in hand, Alexander dissected a recently deceased dog of similar size (obtained from a vet) and then used precise laboratory instruments to simulate the forces created by Happy's jump on the constituent parts of the cadaver dog's legs. He wanted to know how close this force brought these parts to the breaking point, and how much additional force his simulation would need to surpass it. But to do that, he needed to puzzle out how the disparate parts interacted. The first thing Alexander noticed was that the dog muscle didn't move very much, no matter what you did to it. But when Alexander simulated the force one would expect on the Achilles' tendon, he was shocked at what he discovered.

"At that time any anatomist would have told you that tendon is not extensible, that the tendon is not elastic," he recalls. "But we came to what at the time seemed a rather amazing conclusion."

When the force was applied to Happy's tendon, it stretched by a full 3 centimeters. Alexander would subsequently demonstrate conclusively with an exotic coterie of other animals, including kangaroos, wallabies, and a camel, that it was not the muscles that constituted the leg's springs—as many had predicted—but the tendons.

The Achilles' tendon, in particular, Alexander would conclude, accounted for roughly 35 percent of the energy we use when we run. By studying amputated human feet (from patients with peripheral vascular disease) and camels, Alexander soon identified a second spring, this one in the arch of the foot, which he determined fed another 17 percent of the energy required for each stride into the leg. Together these two tendon springs account for roughly half the energy used in each stride when we run. Alexander had solved Cavagna's mystery.

The muscle's role when it contracts, scientists would later learn, is to serve as kind of a backstop—a wall of resistance that redirects the energy the tendons absorb when the leg hits the ground back downward, causing the tendons to stretch with potential energy, like rubber bands. The stiffer the muscle, the more the tendon stretches, and the more energy is stored.

It was left to a young Harvard graduate student named Norm Heglund to extend these findings to bigger, less docile animals. Soon after Alexander's influential paper on kangaroos, Heglund was assigned the unenviable task of banging on a pot lid with a stick and shouting at the top of his lungs to try to scare, coerce, and entice a virtual Noah's Ark of larger animals up and down a hallway in the basement of Cavagna's laboratory at the University of Milan. Watching safely from a distance, a team of more senior scientists recorded the results with video equipment and force plates. Heglund's subjects included two stump-tailed macaque monkeys, a wild turkey, a spring hare, a couple of dogs, a ram weighing roughly 185 pounds, and a kangaroo.

"The monkeys," Heglund recalls years later, "were the worst because they were smart. When they get tired of repeating the experiment, they would start by screaming and yelling and running all over the place doing everything except what you want them to do. Then they would start to defecate in the hands and sling the excrement at you. Finally, they will simply physically attack you with their teeth, fingernails, whatever."

In the end Heglund, Cavagna, and C. Richard Taylor, a Harvard biologist in charge of the famed Concord Field Station, identified two distinct categories of locomotion and energy-efficient techniques. The first explained running, the second walking.

When we run, the springlike tendons stretch as the leg impacts the ground, changing shape to store elastic, or mechanical, potential energy. When our foot leaves the ground, the tendons release the stored energy like a rubber band, propelling us upward and forward, and into our next stride. Essentially, explains Heglund, we bounce along, like a basketball or a pogo stick.

Our calf muscles, meanwhile, shorten and lengthen primarily to modulate stiffness, thus effectively serving as a volume switch for the Achilles' tendons. The stiffer the muscle, the more it pulls on the tendon, which affects how tightly the tendon will coil. (Consider, for instance, how we deal with a puddle we might encounter while jogging. To modulate our step, we flex, stiffening the calf muscles and bringing the tendons in tight, to allow us to take a stutter step and bring us up short. Then we release all that stored energy to vault over the water.)

The team also delineated a second category of movement for walking. If when we run we bounce like a basketball, the way our body conserves energy when we walk is more comparable to the swinging of a pendulum, another man-made device ingeniously designed to conserve and recycle energy. To be more precise, when we walk, the body is like an inverted pendulum—the trunk of the body serves as the weight on the end of the string, and the leg plays the role of the string itself. Like a pendulum, as we move, the body's centers of mass go up and down, accelerating and decelerating within each step. The leg, meanwhile, also alternates, expending energy and losing speed as it pushes upward against gravity, and regaining energy, momentum and speed on the way back down, as the same gravitational forces that slowed it on the way up now propel it forward. This gravity-driven, recycled oomph, Cavagna estimated, accounted for up to 60–65 percent of the propulsion driving each stride when we walk, leaving only 35–40 percent to be supplied by the muscles.

The work of Cavagna, Taylor, and Heglund gave a scientific explanation to what Hugh Herr's antiquated prostheses sorely lacked. In the normal leg the tendons and muscles of the body form a delicate web capable of juggling energy back and forth, storing it, and releasing it. When Herr walked with his lifeless prostheses, there were no tendons or muscles to capture and recycle the energy; they were just dead weight. Sure enough, one day in the not-too-distant future, this realization would prove fundamental to the efforts of Herr and others to revolutionize the field of prosthetic design.

But upon learning the basics of biomechanics, initially Herr had another question: Was there a way he might use these insights to improve his performance on the climbing wall?

On a clear day a couple of years into his graduate studies at MIT, Herr made his way to the foot of Colorado's famed Eldorado Canyon, just outside Boulder in the craggy foothills of the Rockies. He was on vacation and dressed in a skintight black Lycra bodysuit, his thighs balanced on a pair of stumpy metal rods attached to tiny baby-sized feet. Herr's most eye-catching sartorial choice, however, snaked out of the fluorescent yellow climbing harness hanging around his waist.

Instead of the standard safety ropes and metal clips favored by most climbers, Herr used the harness to anchor strands of a long elastic material resembling braided rubber bands that attached to the underside of his upper arms. He called it his "Spider-Man suit." In case anyone missed the superhero/daredevil theme, what Herr did next drove it home: He began climbing the wall without any safety ropes.

Every time Herr reached upward for a new handhold, the web of rubber springs connecting his triceps to the harness drew taut, like a set of synthetic tendons, forcing him to push through the resistance, recruiting his triceps and back muscles into the effort. Webbing also provided extra resistance to the fingers when he opened his hands, as he reached up to grab a handhold. All of this potential energy was stored in his Spider-Man suit, thanks to the artificial tendons drawing energy out of muscle groups that would normally sit idle during Herr's climb.

Next, as Herr pulled his body weight up the wall using a

different set of muscles, the elastic webbing would slowly release the stored energy, helping hoist him upward and halving the strain on his shoulders and biceps muscles. Soon Herr was six stories up.

In a recent video, you can see his able-bodied climbing partner far below struggling to keep pace as Herr reaches the top and pumps his fist in victory. He still had it. And with technology, he was pushing it even further.

"Can you get more work out of the body before it fatigues by attaching a mechanism to the body?" Herr asks. "That was my question. The answer is yes. You are effectively doubling muscle mass to do the same workload, and you can dramatically delay the onset of fatigue. Stated simply, you can make people twice as strong."

Herr had another question inspired by his new knowledge. Could he use what was known about the springs of the human body—and other animals—to augment running speed? To find out, he began designing a running shoe. The shoe contained two springs, one in the heel and one in the toe. Herr connected them with a strip of carbon running the length of the shoe. When a person's heel hit the ground, it depressed the heel spring, storing potential energy. As the foot rolled forward, gradually shifting weight, the heel spring's potential energy traveled beneath the foot's point of contact with the ground, until it reached the toe. Then, when the runner lifted his toe from the ground, the toe spring released that energy—catapulting the runner forward with the extra force. Herr experimented until he had identified the optimal spring placement for energy amplification. Not only could it increase speed and reduce the metabolic cost of running; it could also soften the force exerted on the joints by as much as 20 percent.

Herr offered the shoe to Nike, which took him seriously enough to hire Harvard's Thomas McMahon, one of the leading experts on biomechanics at the time, to evaluate it. Though the company ended up passing on the product, McMahon was impressed. And suddenly Herr had the ideal mentor to take his creations to the next level. In 1990, McMahon had detailed a seminal theoretical and mathematical framework that took the profoundly complicated dynamics of human locomotion and reduced them to simple equations that allowed one to make reliable predictions about movement.

McMahon encouraged Herr to enroll in his class at Harvard. Eventually he would become Herr's Ph.D. thesis advisor. Rather than thinking about all the disparate joints, muscles, tendons, and ligaments of the leg separately, McMahon suggested the entire limb could be considered a single spring. Using this approach, the Achilles' tendon and the springs in the arch of the foot could be considered just a couple of links in one big, bouncy mechanism. This approach worked because, like a single spring, the force of a limb could be compressed to varying degrees depending on another simplified variable: the total amount of weight converging from different parts of the body and exerting downward or outward force on a single point in space. In physics this is called a "point mass."

McMahon showed that if you knew a point mass, and if you knew the angle at which, for instance, the bottom of the leg hit the ground, you could predict how long the leg would spend on a surface before bouncing upward, and how much it would compress. You could determine how much force the leg would explode off the ground with, and how one's center of mass would move through the air between strides. Conversely, if you measured how long the leg remained on the ground between strides, you could, with other variables, calculate the point mass.

Under McMahon, Herr spent months puzzling out the elegant and bewitching mechanics of horses in full gallop. With all four of their legs seemingly simultaneously airborne at times, horses come closer perhaps than any other quadruped to flying, because of their biomechanics. Yet the biomechanics underlying this miracle of locomotion remained a mystery. How did they manage to remain stable? Herr came to believe horses use their legs like compliant springs, perfectly calibrated to provide the ideal stiffness to promote both stability and speed, a delicate balance that hits the exact sweet spot maximizing airtime, while still allowing the animals to remain in control. Meticulously he built a mathematical model that could express the solution to the mystery and capture the grace of their movement.

Herr would earn is Ph.D., modeling the dynamics of a whole host of quadruped animals, including elephants, mice, and everything in between. But as he did the work, Herr began to consider a more ambitious project—one that many people would have told him at the time just wasn't possible. For years Herr had been forced to rely on stiff, awkward prosthetics that didn't come close to matching the dynamism, power, and ease once provided by the legs he had been born with. He'd had to go to the climbing wall to taste true freedom of movement. Now Herr began to consider how he might build a better device. He wanted an artificial limb that would allow him to walk almost as if he were on natural human legs again.

Hugh Herr rises from a chair in his glass-walled office on the third floor of the MIT Media Lab and leads me down a thin walkway overlooking a vast open workspace. Gripping the metal banister of a spiral staircase, Herr descends methodically and apparently effortlessly on the pair of mechanical legs he built.

Soon we are standing in the well of a spacious laboratory, a mechanical magician's workshop full of stacked toolboxes, long desks arrayed with hammers, wires, and drills, and any number of cubicles for the small army of graduate students and aspiring engineers who work for him. Thickets of overhanging wires curl off the edges of tables and desks, disappearing to unseen machinery and motors behind metal drawers and file cabinets, like overgrown vegetation sprouting from the walls of a jungle fort. If clutter is a sign of creativity, ideas aren't a problem here.

We have arrived at the heart of an ambitious effort led by Herr to unlock the mysteries of human movement and use that knowledge to build bionic body parts capable of replicating, and in some cases surpassing, it.

I follow Herr over to his new 3-D printer, which he tells me he plans to use to print out prosthetic parts. Then we walk past desks littered with stray prosthetic arms and legs, and monitors. Finally we stop in front of one of the room's most prominent and eccentric features: a long, slightly elevated treadmill—resembling in shape a sizable portion of one of those moving walkways one finds in airports. Hung from the ceiling and arrayed around the treadmill, facing downward toward the track at various angles are more than thirty cameras that Herr and his team have installed.

Before instructing one of his experimental subjects to step onto the elevated treadmill—or before stepping onto it himself—Herr affixes scores of 1-centimeter-wide reflective markers at well-defined anatomical locations across the body. Then when the subject, or Herr, steps onto the treadmill and begins to walk, all anyone needs to do is press a few buttons and the cameras will begin gathering precise data about how the constituent parts of the human leg are interacting to produce movement, tracking the marker positions as they move through space, and conveying the information to a computer where it can be analyzed.

This data allows Herr and his colleagues to determine exactly how the knee angle, for instance, changes over time. The way that the motion of the right thigh reflects changes up and down the ankle. Or how that relates to the inflection of the foot.

Motion capture systems, available today perhaps most famously from a company called Vicon, have revolutionized not just the way engineers like Herr study movement in recent years but a wide array of other disciplines. Animators use them to record real actors and make their characters move on the screen in a lifelike manner.* Maybe you've seen LeBron James in one of those commercials from the video game manufacturer EA Sports, covered in little reflective balls slamming a basketball home? That's how EA's animators make his video game doppelganger credible. But this technology is not just improving fantasy sports. Trainers working with the Boston Red Sox, the San Francisco Giants, and the Milwaukee Brewers use the technology to record the throwing motions of their pitchers—and then suggest tweaks to their form that maximize the fluidity of the motion and the force that it can generate. And in a laboratory at Southern Methodist University, in Dallas, a biomechanics professor named Peter Weyand has worked with some of the best sprinters in the world, analyzing the mechanics of their leg movements in his lab and on video, both to understand what makes them so fast and to suggest tweaks that could optimize their motion.

By using motion capture and computer analytics, Weyand has demonstrated, among other things, that the amount of speed top sprinters generate is related to the force and timing with which their feet hit the ground—which allows them to bounce greater distances. This speed bears little relation to the isometric strength of these sprinters—how much weight, in other words, they can push upward with their legs.† Rather their speed is related to the timing

---

* The British actor Andy Serkis has perhaps most famously used motion capture to portray computer-generated characters such as Gollum in *Lord of the Rings,* a giant ape in *King Kong,* and the Supreme Leader Snoke in *Star Wars: Episode VII; The Force Awakens.* James Cameron used the technology to great effect in the 2009 movie *Avatar.*

† Runners like Usain Bolt, Weyand notes, hit the ground with force one and a half to two times that of normal runners, four to five times their body weight. They can do it in three to four one-hundredths of a second of their first contact with the ground, far faster than anyone else, says Weyand. When Weyand slowed down footage of sprinters like Bolt and Carl Lewis, he discovered that their running motion is geared toward driving their limbs into the ground with uncommon force, yet at the same time lifting

of their gait and the angle and force with which their feet hit the ground and linger there—factors that can be maximized by proper form, and perfected with practice.

Herr has another use for the technology. When he began designing a human leg after earning his Ph.D., virtually all commercially available ankle and foot prostheses were passive devices. The designers had built in spring mechanisms to absorb shock as a person walked but made no effort to replace the power-generating capabilities present in the muscles of individuals who still had the lower limbs they were born with. To Herr, this design decision seemed to guarantee problems. Thus he decided the ankle and foot were the obvious place to begin.

Herr had carefully studied the work of another disciple of McMahon, Claire Farley, who had demonstrated beyond a shadow of a doubt in the 1990s that the human ankle is, in fact, the primary joint we use to modulate the stiffness of the entire leg. Since greater stiffness is what allows the leg to produce greater bounce—and more power output when it is needed—Herr knew that the ankle could even be considered the leg's *primary* "motor." By varying the amount of muscle activation and thus stiffness and bounce, the ankle acts as a volume knob to turn up or down the power and the speed with which we walk.

"The changes in the ankle joint parallel overall leg stiffness," says Dan Ferris, a biomechanics professor at the University of Michigan, former Farley Ph.D. student, and coauthor with her on several key papers on the biomechanics of the leg and ankle. "What the ankle does drives the whole leg."

To Herr, it seemed clear that the passive, dead weight of pros-

---

them off the ground as quickly as possible. (It makes sense if you watch a race closely. Sprinters stand out for a characteristic gait—witness the perfectly erect body posture, the undeniably high knee lifts, the way the legs pump up and down, blindingly, rhythmically like pistons.)

thetic ankles could explain any number of the ills experienced by lower-limb amputees. Even with the best available models, most amputees walked more slowly and had less balance. Their gaits were eccentric, and their devices often caused back problems. Perhaps most important, when a person with intact lower limbs walks, the amount of power the calf muscles expend increases with walking speed. Herr believed that the lack of ankle power in prostheses was one of the main reasons amputees burn 30 percent more energy walking than do humans with intact lower limbs. Without a functioning ankle to modulate stiffness and bounce, the walking was far more inefficient.

"I started to think about prostheses I would wear, how important it would be for a prosthesis under computer control to vary stiffness when a person walks, when a person runs," Herr recalls.

Herr set out to create a mathematical model that spelled out precisely how the different components of the lower leg interact. In order to do so, Herr had to ask fundamental questions about everyday behavior. How much power, for instance, does a normal calf muscle in a 5-foot, 9-inch male generate right before the foot pushes off the ground? When that muscle flexes, how will it affect the stiffness of the tendons attached to it? How stiff is the ankle when a person attempts to slow down?

To get the data he would need to answer these questions, Herr and his team spent months rummaging through reams of previous scientific studies and collecting all that was known about the dynamics of the human leg and the interaction of its component structures. Where the literature was sparse, Herr tried to fill in the blanks by recruiting able-bodied volunteers and using motion capture technologies to characterize their movements.

While creating this mammoth mathematical description of a leg at work, Herr began designing a robotic prosthesis capable of translating this math back into movement. To replicate the ankle's natural ability to brake when walking downhill, he modified a previous invention he had created to control the stiffness of a prosthetic knee. The device contains sliding steel plates separated by an oily liquid that grows thicker when a magnetic field is applied. Electrical sensors measure both the angle and the force applied by the user on the ankle, and a computer modifies the strength of the magnetic field accordingly. Then, to determine the ankle's location in space and to adjust the angle of the prosthetic foot appropriately (if, say, a person's foot is suspended in midair going down stairs), Herr incorporated the same sensors used in guided missile systems.

To track his progress, Herr created a digital avatar, which he shows me, displayed on a large monitor.

It's a crude, digital representation of a torso with legs, which walks as if drunk, or blind, across the screen. Though the graphics are basic, the cartoon figure's lower limbs are actually composed of hundreds of virtual tendons, muscles, and bones, each programmed to emulate different parts of an actual human leg. How much joint torque is applied at the ankle or the knee? How much electrical activity is occurring at the muscle? How and when do the tendons in the leg capture and release energy? The cartoon captures all that data and displays it on the screen as a virtual representation of how the body would walk, if perhaps blindfolded, obeying the laws of motion.

The same mathematical descriptions used to determine how the cartoon walks are programmed into the software controlling the movement of the constituent parts of the ankle-foot prostheses that Herr wears on the stumps of his legs this very day.

It's amazing to contemplate the idea that, as I stand next to Herr, the tiny chips buried somewhere in the machinery hidden behind his pant legs are able to perform unimaginably complex computations every second, to guide the behavior of every single part of Herr's bionic limbs. Herr derived these formulas from real-world measurements and observations about not just the way the parts of real human limbs behave in isolation, but also how they interact with one another. How stiff the mechanical ankle joint is at any given moment might depend, for instance, on how much force the motors emulating the human calf muscle in the prosthesis are exerting on the gears meant to emulate an Achilles' tendon. But it also might be affected by which way the knee joint is torqued and at what angle, taking into account, perhaps, the speed at which the upper leg is moving forward or down. There are, in short, a dizzying number of factors in play at any given moment.

Yet Herr's computer program doesn't tell the entire bionic leg how to move. It's not a "playback machine" with preset movements, he likes to say.

"Playback machines don't work," he says. "What happens when you step on a banana peel?"

Instead, the electronic package meticulously programmed by Herr and his team tells each *individual* part of the bionic limb how to respond to the many different kinds of inputs around it—from the amount of pull exerted by the artificial tendons, to the angle of the ligaments and the force of muscles that surround it. Like the natural leg, Herr's robotic limb is a dynamic collaboration of many different parts, pulling and pushing against one another, bending and stretching and recoiling. The result, he explains, are qualities and behavior that sometimes surprise even him—behaviors that are "emergent."

"We're not telling the model how it moves," he says. "The model tells us how it moves."

"We take actual measurements from our sensors on the prostheses and they are input into the model and the model tells us how stiff and how powerful the joint should be in time," Herr says. "The behavior of the physical prosthesis is then dictated by that mathematical description in the body. It behaves as if it has muscles and tendons, even though it is made of aluminum and silicone and carbon. Even if it is made of synthetic constructs it behaves as if it is flesh and bone."

As miraculous as it seemed, however, it was not gathering the data that proved the biggest obstacle. It was finding a way to power the robots autonomously. Herr's original prototypes were connected to a backpack containing almost thirteen pounds of electronics that amplified power coming from a wall socket—not a practical solution for an amputee on the go. Herr's graduate students spent months trying to reduce transmission losses and cut back on energy costs but could not come close to creating a motorized ankle small enough and powerful enough to match a real one.

Herr found the solution by taking a page from one of the earliest characters in the literature of locomotion—the flea,* with its unparalleled catapult mechanism. In the 1960s, scientists demonstrated that the flea is able to accelerate at a velocity 100 times greater than anything a muscle is capable of spontaneously generating. To achieve this trick, the flea gradually feeds energy into fibrous, springlike structures attached to the muscle and stores it until the moment of its explosive liftoff. Then it releases this stored energy all at once, creating an unbelievably powerful catapult that would put to shame the kind deployed by medieval knights when they besieged cities.

---

* It was British scientist Henry Bennet-Clark who, in the late 1960s, first identified and characterized the tiny insect's remarkable leaping ability. Bennet-Clark and a colleague placed fleas in a tiny cell bathed in an array of jump-inducing heat lamps, and set up what was then considered a high-speed camera. Through painstaking trial and error, they managed to capture twenty "usable" jumps that could be discerned on the tens of thousands of frames of negatives they collected documenting just five seconds' worth of activity. About one-tenth of a second before each jump, the flea gathers its hind legs until its upper quarters are nearly vertical, Bennet-Clark found. Then it remains stationary for approximately eight one-hundredths of a second more, before folding the rest of its legs back, pushing its trunk upward, and hurling itself through the air. Even at a film speed capable of capturing movement every one millisecond, the insect left the ground with such explosive force, its subsequent leg movements were far too rapid to capture. By measuring the length of blurs, however, Bennet-Clark was able to estimate that the flea was accelerating at a velocity that would require a force far surpassing (by about a hundredfold) anything a muscle was capable of generating in such a short period of time.

Herr knew that a small motor alone would not be capable of delivering adequate energy quickly enough to replicate the burst of force with which the foot pushes off the ground when we walk—any more than the muscles of a flea can generate enough power to catapult it from a dog's tail to its back. But if a motor in Herr's artificial leg gradually fed energy into a spring like the flea fed energy into its legs, Herr realized, the speed of energy production wouldn't matter. When it came time to push off, that spring could release all that pent-up energy at once, propelling the human foot off the ground with the explosive force of a natural human ankle.

Herr's lead graduate student on the project, Samuel Au, spent months tinkering unsuccessfully with the motor. Then Herr realized that none of the versions of the motor incorporated the secondary usage of tendons that happens in a real ankle joint. Perhaps the solution was to add more springs, this time in parallel with the motor.

The hunch paid off. The secondary springs reduced the amount of force required by the motor, mimicking the calf muscle's usage of the Achilles' tendon, which allows the calf muscle to provide power without contracting. To test it out one day, Herr put on the reworked prototype and began shambling down a walkway in the lab. As he walked, a broad smile began to spread across his face. Herr picked up his pace, walking faster and faster. By the time he announced that the ankle felt "just the same as walking with a normal ankle," the lab assistants were cheering.

Making full use of this newly reconfigured web of springs, the lab soon doubled the power emanating from the battery in the small motors in the prostheses. Today when Herr walks, a motor in the back of each foot gradually feeds energy into a combination of springs inside the foot. Some of the energy is released when he simply pushes off the ground. If he climbs a hill or picks up his pace, the motor and the springs release more energy, as required.

"That," Herr says, "is just how the body works."

Herr often dons an oxygen mask in the lab, straps on prostheses, and gets on a treadmill to test upgrades. In addition to tracking the movements of various parts of his body with the Vicon system, Herr might use electrodes attached to muscles around the body to detect the electrical potential generated in muscle cells and measure muscle activation levels, a technique called "electromyography" (EMG). Embedded in the floor itself are two-by-four-foot-wide weight-sensing plates that precisely measure the force the human puts on the ground when he or she walks, dances, or runs—or ground reaction force.

"I am an amazing experimental model," Herr says, "because if you put robots below me that are fully programmable we can actually test hypotheses. If my body responds in a normal way, as if I had intact lower limbs, then that would suggest our theory is robust and is supported. If my body responds in a pathological way, such as consuming far more energy than normal, it suggests our theory needs work."

But the more powerful evidence of the device's utility is revealed in the response of some of the subjects from outside the lab who have tested out the ankle, and their loved ones watching them walk. Often they start to cry.

"It can be very emotional," Herr reflects. "It can feel like you have your biological foot back."

Hugh Herr's achievements allude to the quiet revolution under way in bioengineering and drive home the idea that we are passing the realm of the theoretical and entering an era of transformative real-world impact. In a very real sense, Herr is reverse-engineering parts of intact human bodies. Then he is harnessing technology to emulate the very parts he lost through circumstance. The results are changing not just his own daily life but the lives of countless others like him—men, women, and children who weep with joy when they feel that sublime bounce in their step, made possible by Herr's bionic limbs. It's deeply inspiring.

But, of course, during our time together I can't help but to start to wonder how far Herr might push this technology.

After all, as a graduate student, Herr talked about Spider-Man suits and tricked-out running shoes. Today he remains an intense man with the athletic build of a gymnast and a grace to match it. He tells me about holidays spent rock climbing in the Italian Dolomites. Hugh Herr the scientist and engineer is so clearly still an athlete today that I start thinking about that fictional crime-fighting television hero from my childhood named Steve Austin. Austin, you might recall, was a former astronaut, catastrophically injured and then rebuilt as part of a secret government project. His price tag sounds quaint now considering inflation: They called him "the Six Million Dollar Man." But the moniker was meant to imply Austin was a sleek machine. Austin had bionic vision, could run as fast as a car, could knock down a house with the well-aimed toss of a rock.

Would Herr ever consider not just restoring ability to those who had lost it, but making them stronger than their "able-bodied" peers?

In fact, the days when such a prospect were pure fantasy have long since passed. Not long before my visit, Herr himself was one of several scientific experts tapped to serve on a panel to determine whether to allow the paraplegic South African sprinter Oscar Pistorius to compete against able-bodied runners in the Olympics. The South African sprinter ran on J-shaped, carbon-composite "Cheetah legs" that produced elastic recoil energy every time he hit the ground. Some critics, including SMU's Peter Weyand, claimed the legs gave Pistorius an unfair advantage, because their light weight made them easier to kick through the air between strides. Herr, on the other hand, took the position that Pistorius should be allowed to compete, arguing that the limitations caused by what he lacked more than negated any advantage. Herr's side won.

In the wake of Pistorius's sensationalistic 2013 late-night fatal shooting of his supermodel girlfriend, the world has long since forgotten this battle—and Pistorius's forgettable performance in the actual race. But this does not diminish its import. Weyand has already gone on record predicting that it's all but inevitable that a paraplegic will soon break the world speed record.

So during my time with Herr, I ask him the obvious question. Would he someday invent a device that would allow the paralyzed to run faster? And what about the rest of us: Would he ever invent a device that would allow *me* to run faster?

In fact, even as Herr was achieving the dream of building a life-like prosthetic limb, he remained fascinated by the possibility of using technology to upgrade the abilities we all start with. He is still an unapologetic and prominent champion of the idea that technology should—and will someday soon—be used to augment us all. Herr has been at the forefront of an effort to crack what many consider the biggest challenge in biomechanics, a holy grail of sorts for engineers: building an "exoskeleton" that might make us all stronger, or faster. For some, the idea might evoke dystopian images of murderous Robocops, or U.S. soldiers running amok on the battlefield in Tony Stark Iron Man suits. But Herr sees the technological potential in far more practical terms.

"At some point in this century we'll have a class of human mobility machines that augment the biology of the body, augment walking and running," Herr says. "Fifty years from now when you want to go to see your friend across town, you're not going to go in a big metal box with four wheels. You'll just strap on some wild exoskeleton structure and you'll run there."

Herr defines an exoskeleton as a robot that wraps around a limb (which can be either a normal limb or one with pathology) and can restore or increase endurance, speed, or strength. It's an idea that has been around a long time—long before Iron Man made his first appearance in a Marvel comic book back in 1963.

The first mention of an exoskeleton that would reduce the energy expended to walk, run, or carry a heavy load dates back to at least 1890. Herr came across the plans in the U.S. Patent Office submitted by Nicholas Yagn, an inventor employed by the Russian czar. Yagn proposed to build a device that would use springs to help transfer some of the body's weight to the ground with each stride. (There is no record the device was ever built or demonstrated.)

Despite the long history of efforts, despite revolutionary advances in biomechanics and our understanding of how the human body works—despite, even, exciting advances like Herr's prosthetic bionic limb—for many years a practical exoskeleton has eluded engineers. Most of those developed have been too bulky, have required too much power, or have proven too cumbersome. Without the ability to precisely measure and characterize human movement, most engineers have created devices that have simply gotten in the way.

In recent years, however, some engineers have begun to make dramatic strides in adapting the same biomechanical principles Herr uses to build his artificial limbs, which is now widely available. In fact, even the most basic insights about how muscles function can provide powerful effects. In Tokyo, I visited the lab of a Japanese roboticist named Hiroshi Kobayashi, who has built a basic strength augmentation device for the upper body that is among the first of a new class of such devices being marketed to help health-care workers pick up elderly people without damaging their backs. Kobayashi's device, which he calls a "muscle suit," consists of a backpack-like aluminum frame equipped with four artificial "muscles" made of mesh-encased rubber bladders attached to heavy-duty wire. When compressed air is pumped in and out of the bladders, they change shape like muscles, pulling up the wires attached to pulleys, which in turn shorten the frame, activate artificial "joints," and pull up the person wearing it. The motion offers a powerful burst of strength to augment a wearer's normal back muscles.

When I strapped on the muscle suit, the shiny aluminum device felt lightweight on my back, not much heavier than my empty gym bag. It was so light, in fact, that as I bent down to grasp a milk crate containing more than ninety pounds of bagged rice, I was dubious the backpack would do much for me. But Kobayashi pressed a button, there was a whoosh of air, and I instantly straightened up without thinking about it—slightly disoriented. I had just hefted a load that would have been certain to throw out my back, and I had held the box with my fingertips. It felt like I had reached down and picked up a piece of paper.

Of course, Kobayashi's muscle suit has a downside. It relies on compressed air—and air-compressing machines are very heavy and very loud, like a vacuum. If I wanted to impress my wife and friends at a party by, say, picking up heavy boulders or flipping over cars, I'd probably need some sort of specialized luggage cart to tow around a large supply of the stuff. It's a pretty inadequate machine when placed alongside the human body.

But the lightweight aluminum backpack still provided a tantalizing glimpse of what may come in the future—and how freeing it might actually feel. Though Kobayashi's muscle suit consists of a simple synthetic muscle, moving in a single direction, a number of researchers are making fast progress on far more complex and lightweight devices that more accurately emulate a more complex web of muscles and tendons, and that might soon replicate and exceed the force and dynamic motions of a human arm.

The model of the human hand and arm that Patrick van der Smagt helped develop, while not an exoskeleton, is complete with motors and parts, inspired by muscles, tendons, and bone structure, that the team at the German Aerospace Center hopes will soon offer its wearer far more upper-body strength than anything seen in the natural world.

"Ten years ago, we couldn't even build arms that behaved similar to human arms—far from it," Van der Smagt says. "But the quality of the technology has improved radically. Now I have no doubt we will soon build a bionic arm that is far stronger than a human arm."

The most challenging current bottleneck for devices that augment the body remains the same one that so flummoxed Herr and his team when they set out to build his bionic ankle. Though it's relatively easy to build a motor that can exceed the power output of normal human muscles and then feed it into a proportionally accurate bionic arm or leg, building one that is energy efficient and lightweight enough to be practical remains a major challenge. The humanlike arm developed by the team at the German Aerospace Center weighs about twenty pounds, more than twice as heavy as a natural human arm, and thus is far too cumbersome to attach to a human shoulder. In other words, one might actually be able to build an arm as strong as the arm given to Steve Austin, the Six Million Dollar Man. But to give someone an arm with the strength of a bulldozer, you would need a motor close to the size of a bulldozer's motor to power it.

For now, that means most robotic arms and upper prostheses currently use electrical motors, which are small and lightweight but don't come close to matching the energy efficiencies of the human body. The limitation in energy efficiencies is apparent in the most advanced commercial prosthetic arm created to date. In May 2014, the Food and Drug Administration gave final approval to the arm, made by Manchester, New Hampshire–based DEKA Research & Development Corporation, a company founded by Dean Kamen as part of a $100 million research program funded by the Defense Advanced Research Projects Agency (DARPA). Kamen's company aimed to create a device that matched the size and weight of a real human arm, about 7.9 pounds.

DEKA's arm can detect signals from muscles beneath the skin—using the same kind of sensors Van der Smagt used to measure muscle activation from his volunteers—which implies one might do so with an augmentation device, like an upper-body exoskeleton. The

DEKA arm can then react to the muscle activation signals by opening and closing the prosthesis and change grip configurations. With the device, users have been able to pick up a coin, and drink a glass of water.

But to engineer the DEKA arm to the size and weight specifications of a natural human arm, Van der Smagt notes, Kamen and his team had to sacrifice other features, including strength. A normal human arm can generate force to act on weight as much as twenty times its mass. But the DEKA and most other prostheses have a force-to-weight ratio many times smaller, he says. An 8-pound human arm, he notes, can usually pull or push 200 pounds—more than twenty times its weight. Kamen's DEKA arm can't come to close to that.

"The DEKA arm's force-to-weight ratio is not very good compared to the nonprosthetic arms I have worked with and certainly not compared to the human arm," Van der Smagt says. "Also the stability is not as good as you would like it to be. It's not power efficient. They do some very good engineering and it's the right shape and weight and probably the best prosthetic arm around. But it is definitely not biomimetic."

Kind of disappointing when you consider that the price tag was enough to build 16.6 "Six Million Dollar Men." In that sense Steve Austin is still a dream.

In fact, perhaps the most promising device that might allow humans to actually exceed normal performance comes, once again, from Hugh Herr. In 2014, Herr announced he had created the first lower-limb device in history that actually assists an able-bodied individual as he walks while at the same time reducing metabolic cost.

The litmus test for a useful exoskeleton, Herr argued, was whether it was capable of feeding power into an individual stride without actually increasing the metabolic cost to the person wearing it. It was a challenge that no engineer had ever managed to overcome.

In a demonstration video showcasing the new technology, a subject in blue shorts, wearing what look like standard army desert boots and a pair of knee-high black socks, strolls along a treadmill. Strapped to the front of each ankle, a couple of inches below the knee, is a small black object, no bigger than a cigarette box. This little box is the device's "artificial muscle."

A pair of long, thin black metal struts run along either side of each leg, connecting beneath the arch of the foot, then projecting up and backward, into the air behind the calf, at a steep diagonal. These struts are used to help the motor on the other side of the ankle distribute power and augment the soleus muscle, a long, powerful array of fibers that runs from behind the knee to the heel, attaches to the Achilles' tendon, and plays a key role in powering standing and walking.

The key mechanical innovation for this device, Herr says, was finding a way to feed energy from a motor into the body in a natural way, without breaking the skin or stressing the leg. Herr's solution to the problem is rather elegant and departs from nature: He feeds the mechanical energy into the stride with the perpendicular device—projecting out of the cigarette-pack-sized motor—pressing in on the upper ankle, which eliminates any rubbing against the skin and the risk of skin shear.

The mechanical power is applied as "torque"—force that helps push the front of the ankle back. This swings open the ankle joint connecting the lower leg to the foot, like the hinge of a door. That movement in turn pulls on the tendons, which pull up the heel of the foot, seesawing it up and pushing the foot and toes against the ground and storing potential energy. When the user lifts the foot off the ground, the power is released, slingshotting the wearer forward.

Herr's exoskeleton incorporates an ingenious feedback mechanism taken directly from nature (a mechanism that he also uses in his prosthetic lower limbs). This mechanism allows the exoskeletal motor to adjust in real time to the changing terrain and speed.

While building the mathematical model for the ankle-foot prostheses that he wears to power his own walking, Herr stumbled on a number of powerful so-called emergent properties—ways of doing things that make perfect sense, but which he couldn't have predicted. One of the most powerful had to do with how the human body feeds additional energy into the lower limb when we walk over uneven ground.

Sitting at the junction between the Achilles' tendon and the soleus, as at other muscle-tendon junctions, is a structure called the "Golgi tendon organ." The Golgi organ is a biological sensor that responds to force exerted onto it by sending a signal up the spinal cord to the brain. The brain responds by instructing the muscle to contract further, increasing the stiffness and power of the leg. When Herr included this structure in his mathematical model of his leg, he discovered it played a crucial role in walking.

"It's very, very simple, and we included it in the ankle prosthesis and it has this amazing emergent behavior," Herr says.

As the amputee—or someone wearing the exoskeleton—goes from slow to fast walking, the pressure on the Golgi organ increases, and thus the model tells the motor to give the ankle more power.

"It happens automatically, without any direct measure of walking speed," Herr says. "Also, when the ground increases, when a person starts walking up a hill, it gives even more energy. When a person goes downhill, it actually takes energy out—automatically, even though it has no sensing for ground change. This very simple muscle reflex has these emergent behaviors that are extremely powerful.

"I would argue," adds Herr, "that [if] you take a genius in engineering and give him or her every class in engineering control theory, they probably wouldn't come up with these simple reflexes."

All of this may sound straightforward—nature's solutions usually have an elegance to them. But the results will likely prove transformative. By using this device, Herr claims, he has created a boot that will allow subjects to walk utilizing 20 percent less energy.

"There's only been one active exoskeleton in history that works," Herr claims. "And that was ours."

In principle, this energy savings could remain true, with some modifications, if a person were to wear a heavy backpack or run really fast. When a person carries a load, Herr notes, it's primarily the knee and the ankle that are forced to change biomechanically, using muscle power to counter the force the load is exerting downward, to balance out the torque.

"You could have an exoskeleton around the knee and ankle that does what the body does, when you carry a load," he says. "But the human inside is walking as if they're not carrying the load."

As I leave Herr's lab, it's hard to separate out the therapeutic potential from the potential for human augmentation—though, at this early stage in my journey, that simply evokes for me intriguing images of real-life Iron Man suits, and the idea I might someday have a machine that allows me to pick up a car. It's an experience I will have repeatedly. Again and again I find examples of technologies that will both restore lost function and facilitate different kinds of human augmentation.

Certainly the therapeutic is the most inspiring. During one of my visits, I ask Herr about that dream that used to torment him when he first lost his legs—the one where he is running through the fields behind his house, the wind in his hair. Does he still have it? No. Hugh Herr no longer needs the dream. Virtually every day for a number of years now, Herr tells me, he has been jogging the 1.7-mile wooded loop around Walden Pond in Concord on specially designed prostheses.

"I was out just yesterday," he says. "It's a beautiful run."

CHAPTER 2

THE BIRTH OF BAMM-BAMM

DECODING THE GENOME AND REWRITING IT

Hugh Herr is able to create lifelike, bionic prostheses and exoskeletons because of the new technologies that allow him and those who study biomechanics to precisely record the way different parts of the body move and interact, and then build complex arrays of robotic parts outside the body capable of emulating them in real time. This requires an amazing amount of instantaneous sensing and processing power, both to capture and characterize the behavior of a healthy leg and to build a machine that can imitate it.

But these feats are just a bare beginning of what is possible. As we will see in the chapters ahead, the same technological precision roboticists are harnessing to build devices that attach to the outside of the body, the same mathematical wizardry and pattern recognition software Herr is using to power his creations—all of this technology can also be trained inward to record, characterize, and understand the way different parts of the body work together on the cellular level. Here too scientists are discovering and unleashing latent healing power and untapped potential that scientists of previous generations could only dream about.

What they are accomplishing is, in some ways, even more astounding than what is transpiring in Herr's lab. Some scientists aren't just building new body parts, or upgrading the ones we have; they are hacking into the body itself and rewriting or redirecting the body's cellular instruction manuals. And by doing so they are coercing the body to rebuild or transform itself. The ideas for some of these technological feats, like the inspiration for Herr's amazingly adaptive bionic limbs, aren't wholly the product of human imagination. Almost invariably, the best of them also come from nature itself.

Consider, for instance, the case of an amazing little boy from Muskegon, Michigan, named Liam Hoekstra.

It was sometime in the winter of 2005 that Dana and Neil Hoekstra first knew for sure that their adopted son Liam was different. That was the day their cheerful, dark-haired baby, all of five months old, reached up to grab his mother's two extended forefingers. Holding them in an iron grip, he then hoisted himself straight off the ground and stretched out his arms to form a human T in midair.

His parents had seen Olympic gymnasts perform the move on television to demonstrate their uncommon strength. It was called the "Iron Cross."

"He would, literally, just hang there," says Neil.

By three, Liam had six-pack abs and bulging biceps. He could scale a gym rope unassisted and was waving around five-pound dumbbells as if they were toy rattles like Barney Rubble's terrible tot Bamm-Bamm on *The Flintstones*. One day he had a tantrum—and punched a hole in the wall.

It wasn't until Liam's grandfather, a retired attorney, bragged to a doctor friend about how his diminutive grandson would one day play football for his beloved Michigan Wolverines that the family finally learned the likely cause of his uncommon strength. The doctor asked to examine the boy himself, then convinced Liam's parents to send him to a specialist for genetic testing in nearby Grand Rapids, Michigan, who sent Liam's genetic samples on to the University of Pittsburgh.

Researchers there informed the family that Liam's remarkable abilities were likely the result of a single mutation—the equivalent of a single typo—somewhere in the three-billion-base-pair-long genetic sequence encoded in every one of his cells.

"We're assuming he has a mutation because he has such an unusual hypermuscular phenotype," says Robert Ferrell, then codirector of the University of Pittsburgh Genomics and Proteomics Core Laboratory. "We just haven't found it yet."

Ferrell believes Liam's mutation lies close to the site of a mutation found in another baby and detailed the year before Liam's birth in the *New England Journal of Medicine.* The unnamed German subject in the article has a mutation in the gene GDF-8 that makes him unable to produce myostatin, a signaling agent that plays a key role in regulating and limiting muscle growth. A similar mutation, somewhere in the same biological pathway, would explain why Liam has 40 percent more muscle mass than average children his age, eats six meals a day, and can bench-press the family dog.

It would also bode well for Liam's future chances with his grandfather's beloved Michigan Wolverines. Though the identity of the German boy has never been revealed, this much is known: His mother, who also has the defective gene, is a professional sprinter. Her grandfather was a construction worker who could lift curbstones weighing hundreds of pounds with his bare hands.

The discovery of individuals like Liam with uncommon, potentially superhuman physical characteristics has taken on new meaning as we enter the age of genetic engineering. People of uncommon strength, flexibility, height, and endurance, of course, appear throughout recorded human history—from Hercules to the strongman with the handlebar mustache, shaved head, and leopard-print one-piece at the traveling carnival.

But new technologies suggest we might soon use what we learn from superstrong individuals to treat and possibly even cure some of the most devastating genetic diseases of our time. These same technologies, however, also raise complicated questions. What happens when all of us have the option of giving ourselves or our children the strength of a Liam Hoekstra, permanently? If we choose not to, will we be condemning our children to a lifetime in which they will lose out to genetically modified competitors whose parents reach a different decision?

I'm on the New Jersey Turnpike staring out at smoking, smelly chemical factories when I get the idea that I should listen to some sports radio on the car stereo to get in the mood for my upcoming interview. It isn't until I enter the outskirts of the City of Brotherly Love, however, that I really start to pay attention.

The radio discussion about the hometown football team, the Philadelphia Eagles, has turned emotional, as if the quarterback, who has been struggling, were a personal friend with a chronic drinking problem or an abusive spouse. The callers express all the stages of grief: anger ("We should cut him loose. We're enabling him."), denial ("It's only temporary"), bargaining ("If we get a good receiver, he'll do better"), sadness ("I can't take it anymore").

As I pull into an underground garage on the gritty, urban campus of the University of Pennsylvania, I reflect on our bizarre infatuation with a game that involves three-hundred-pound men in tight pants and pads running around colliding with one another. For some of us, winning at football really can come to seem a matter of grave, existential importance.

That's something the man I'm on my way to meet, H. Lee Sweeney, experienced firsthand in the most head-shaking, bizarre manner. Back in the late 1990s, Sweeney pulled off an extraordinary scientific feat. He had created one of the world's first genetically engineered supermice, a regular-sized, unassuming lab rodent that he had magically transformed into a specimen with legs so muscular, so absurdly pumped, that the press wasted little time dreaming up a sensationalistic name. They called the mouse and his siblings "Arnold Schwarzenegger mice."

Sweeney told an electrified crowd at the American Society for Cell Biology conference in San Francisco that his technique could one day help senior citizens whose muscles were wasting away, or buy some time for patients afflicted with the deadliest forms of muscular dystrophy. It was an inspiring vision that promised new hope for an intractable disorder that had seen too little.

And it's true that upon his return to his laboratory, Sweeney was besieged with phone calls from desperate patients and the loved ones of some of the weakest among us. But Sweeney's phone lines also lit up with calls from athletes, able-bodied men and women in the prime of their lives. These athletes begged Sweeney to try out his experimental technique on them.

"The calls and e-mails started the day that paper came out," Sweeney says. "Hundreds of them."

One high school football coach even offered to pay Sweeney to modify the genes of his entire team. Sweeney, a mild-mannered scientist with an understated air about him, politely declined. But Sweeney's longtime administrative assistant, Barbara Price, was often less diplomatic.

"I was really taken aback a couple times," says Price, who was forced to field the bulk of the calls. "I would say, 'Are you kidding me? Dr. Sweeney's research is with animals!' We even heard from parents of athletes."

Seventeen years after unveiling his first wave of muscular mice, Sweeney remains at the nexus of one of the most ethically fraught scientific conflicts of our times. Unlike Herr, who seems at ease piloting his biomechanical excursions back and forth between the realms of restoration and enhancement, Sweeney is deeply conflicted. He is both fighting to advance genetic engineering and at the same time trying to prevent its misuse. And indeed, Sweeney's chosen area of inquiry is the kind that keeps medical ethicists awake at night.

Today Sweeney is a sought-after speaker at conferences for parents who have children with muscle-wasting conditions and a respected advisor to the World Anti-Doping Agency (WADA), where authorities are wondering when the age of "gene doping" will officially begin—wondering, in fact, whether it has already begun and they just don't know about it.

Sweeney has no illusions. "You could already try to genetically modify an athlete if you have enough scientific knowledge," he says. "WADA does wonder if there are people out there doing gene doping. Some of these power athletes are so obsessed with winning that they'll do anything, even if it compromises them in the long term."

Beefed-up, genetically modified athletes who are able to run over the rest of us Lilliputians with impunity are, of course, just one potential consequence of the burgeoning gene therapy revolution. Any technology that can edit the genes that cause disease also raises the specter of scores of new creations that also make many uncomfortable: armies of genetically engineered supersoldiers immune to pain and incapable of feeling empathy, overbearing parents rewriting their children's DNA to get them into Harvard, designer babies modified to look like Justin Bieber.

Indeed, just as Hugh Herr and his colleagues are finding revolutionary ways to transform the human body with bionics attached to us from the outside, scientists like Sweeney are transforming what is possible from the inside—by digging into the genetic blueprints present in every one of our cells, and adding or altering the details.

Though Sweeney is committed to doing what he can to help

prepare for the possibility of gene doping in athletics, and has heard the concerns about genetic engineering in general, it hasn't stopped his research. There's too much suffering out there, too much potential for healing. That's why in 2011 he took another big step closer to trying his technique out in human subjects: He moved on to a "big animal" model.

Sweeney genetically engineered the world's first "Arnold Schwarzenegger" golden retrievers.

Back when he was in high school, Sweeney played football in two football-obsessed states, Louisiana and Texas. He was the quarterback, which meant that he was the one those three-hundred-pound guys on the other team were trying to crush like a bug.

"I was not interested in bulking up," he says. "I was interested in the other team not bulking up so I could survive."

Maybe that's one of the reasons Sweeney has been unmoved by the pleas of ambitious meatheads begging for help in getting bigger. But as a scientist, he also has trouble relating to their mindset. Science is a slow and plodding pursuit and today Sweeney plays the long game. The able-bodied athletes who contact him, on the other hand, are seemingly willing to risk sacrificing their health in the long term for the chance of a little glory in the now. "Some of these athletes," he says simply, as we sit in a conference room outside his lab, "are crazy."

A calm, modest scientist with a wide, prominent brow and a neat, middle hair part that lends him a distinctly boyish air, Sweeney spent his early career cloistered in the antiseptic, isolationist realms of a laboratory, hemmed in and blocked off from the gritty streets of Philadelphia and the rest of the world by a looming concrete thicket of medical buildings and hospitals and research laboratories. There the white-coated researcher lost himself in a molecular world far removed from the kinds of urgent, real-life human dramas that would eventually come to motivate his research.

From the start, he was one of those lucky scientists blessed by that pure, almost childlike, intellectual wonder that drives the best of us to unlock nature's secrets. It's been like that since the day in the early 1970s, when as an undergrad at MIT—not far from the present-day lab of Hugh Herr—Sweeney hunched over a microscope and first saw a muscle cell in motion.

"It was just so cool that you could see these organizations of molecules actually move," Sweeney recalls. "And you could actually do it with individual filaments of proteins; you could label them and watch them move."

Back then, it was not the extremes of muscles that fascinated Sweeney—children whose muscles seemed to destroy themselves or huge muscle-bound weight lifters intent on seeing how far they could push it—but far more basic questions.

While Herr was driven to measure and replicate how the tendons and muscles of the body capture, juggle, and recycle energy, Lee Sweeney wanted to understand where that initial burst came from to get the movement started in the first place. How did your arm, for instance, go from a posture of complete rest to the blindingly fast movements necessary to hurl a rock? What is the source of the initial burst of energy that can send a sprinter flying off the blocks? How is it that you or I can suddenly pop up out of a chair to shake someone's hand?

Sweeney knew that somehow this mysterious explosion of energy began deep inside our cells themselves. But how could something that started inside a structure so microscopic we can barely see it produce enough force to move a bone? How could it generate the force to allow a two-hundred-pound human to walk, throw a baseball, or turn his head? Beyond that, how did that force even get from these tiny cells to the bone it moved?

Our muscles, Sweeney learned, are composed of bundles of cylindrical fibers no thicker than a strand of human hair. It's these bundles of fibers that you can see in a piece of cooked chicken breast when you pull it and it comes apart in smaller strips. When Sweeney zoomed in on these cylindrical fibers under his microscope, he noticed that these fibers are *themselves* made of smaller strands, called "filaments," which are also knitted together. And if the fibers are like strands of hair, to Sweeney the smaller filaments that compose them resembled tiny threads. The thickest of these threads are made of proteins called "myosin," the thinner ones of "actin."

Amazingly, it's the interaction of millions of these two tiny parts, so small they are hard to see with the naked eye, that allows a 12,000-pound African elephant to thunder across the plains, an NBA basketball player to dunk, and little Liam Hoekstra to pull himself up on his mother's forefingers into a human T.

In the muscles in each cell, thick clusters of myosin strands are laid out parallel to the thinner strands of actin, which are themselves tightly coiled. The ends of the thicker myosin strands in these clusters can curve either up or down, like crooked fingers forming a long line of tips resting on the coils of actin arrayed above and below them. These "myosin heads" form thousands of microscopic bridges to the filaments they are sandwiched between.

At the time that Sweeney entered the field, it was known how the process of muscle contractions started: A decision to move your arm usually begins with a biochemical impulse in the brain—a spike of electrical activity that travels down the spine and out to the peripheral nerves of the body, until it reaches the junction between the nerves and the muscle itself. At that point, the nerves release a chemical called "acetylcholine." The exact molecular mechanisms of the miracle of movement that occurs next had yet to be fully worked out.

What was known is that the chemical reactions that acetylcholine sets in motion cause calcium to be released from storage sites within the muscle itself. This calcium then allows the protein myosin to interact with long filaments of the protein actin. The myosin uses a compound called "adenosine triphosphate," or ATP. ATP is the body's readiest form of stored energy. And, like gasoline in a car or lighter fluid on a fire, it fuels the motion of the muscle. Reacting with the ATP, the muscle's "myosin heads," which Sweeney came to think of as the true "motors" of the body, detach and reattach to the actin, stretching with the elasticity of a bundle of rubber bands, pulling on the actin like grappling hooks and causing the muscle to contract. This we see, for example, as a sudden bulge in the bicep.

The more myosin filaments are bundled together, the harder and faster their grappling-hook tips can pull the actin (because there are more of them), and the bigger the muscle appears to the naked eye.

"Myosin filaments pull on the actin filaments and make them slide," Sweeney says. "That's how muscle shortens."

When Sweeney learned this, he also came to comprehend the biological basis of something we all intuitively understand, but which might seem a contradiction if you stop to think about it: Why is it that the strongest of those among us—those beefed-up linemen who used to chase Sweeney out of the pocket or the members of the Russian women's shot-put team—are the people you are least likely to see running a marathon? Logic might suggest the opposite. After all, if you have more muscles, shouldn't you be able to run longer?

The explanation for this contradiction is simple: There are several different types of muscle fibers. Some fibers are specialized to generate lots of power quickly—the kind you need to fly off the starting blocks at the beginning of a race, pick up a hundred-pound sack of rice like it's a piece of paper, or hurl Lee Sweeney, the high school quarterback, into the turf. And some fibers are less powerful, slower but far more energy efficient—the kind you need to run a marathon, walk into town, or hold your head up all day. The latter are usually called "Type I," or "slow-twitch," muscle fibers. The former are called "Type II," or "fast twitch." (There are actually more than one form of Type I and Type II fibers, but for now let's keep it simple.)

Type II fibers explode with action in a blaze of glory and then quickly burn out when the powder runs dry. They are like the hare in Aesop's Fables who charges out of the gate, then takes a nap in the middle of the race. Type I fibers are the tortoise—slow but steady. They operate on a slow burn, gradually consuming energy as it becomes available, and contracting at a more reasonable and sustainable pace. They can go all day. Given enough time, the tortoise will always beat the hare in a race. Slow-twitch muscles are equipped with more of the cellular machinery that can convert a single molecule of sugar into thirty molecules of ready muscle fuel, in the form of ATP. But it takes more time. When sufficient sugar and oxygen is available, slow-twitch muscles can perform this chemical transformation and power their work without interruption. Their fast-twitch cousins can also produce ATP from sugar—but they can do so much faster. There is a cost for this speed: The process is far less efficient. After an initial burst of energy, instead of 30 molecules, the cruder, more down-and-dirty metabolic processes used by fast-twitch fibers might produce just 2 usable molecules of ATP. They also leave behind chemical waste like lactic acid, which causes that burn we all feel after a hard workout.

An athlete's ratio of slow-twitch to fast-twitch fibers is largely genetically determined and can suggest a predisposition to either sprint or endurance sports. Likewise the animal kingdom, where fast-twitch fibers are dominant in a cheetah's leg muscles, while slow-twitch fibers are found throughout a sloth's legs. But training can also affect the ratio. Olympic sprinters can possess more than 75 percent fast-twitch fibers in their calves, according to some studies, while elite marathoners often have been known to have legs composed of about 80 percent slow-twitch fibers.

All of these insights would eventually prove useful in Sweeney's research. After earning a Ph.D. in biophysics and physiology at Harvard, Sweeney joined the University of Pennsylvania and focused largely on studying the "motors" of the muscle, myosin. In the mid-1980s, however, a team of researchers at Children's Hospital in Boston made a discovery that would broaden Sweeney's research, change the trajectory of his career, and eventually thrust him into the center of the emotional, high-stakes quest to cure a devastating disease.

Often it's studying the things in the body that are broken that tells us the most about what is needed to make things work and why. By the 1980s, Louis M. Kunkel, a professor of pediatrics and genetics, had spent years searching for the genetic correlates of the most extreme form of the muscle-wasting disease muscular dystrophy, a condition known as Duchenne muscular dystrophy. In 1986, Kunkel identified not only the gene where the mutations that caused Duchenne were located, but the protein that gene specified—a protein involved in muscle function that no one had even known existed. Somehow the absence of this protein set in motion the devastating sequence of events that caused the muscles of those with Duchenne muscular dystrophy to waste away.

For Sweeney, Kunkel's discovery of the protein he named "dystrophin" was akin to the discovery of a new planet in the solar system. It opened up a whole new area of exploration. Sweeney set out to solve the mystery of dystrophin's function and began to publish papers about it.

Soon Sweeney began receiving calls from organizers of conferences for parents whose children were afflicted with Duchenne muscular dystrophy.

"Well, I don't work on any therapies," Sweeney told the callers. "I'm just interested in how this protein works and what goes wrong when it isn't there."

"We'd still like you to come talk about it," they'd tell him. "It's important for people to understand more about your insights and how it all works, because that might help them think about how you might fix it."

So Sweeney went. And those conferences changed his life.

If you've ever met anyone with Duchenne muscular dystrophy or someone whose child has it, it's easy to understand the urgency that Lee Sweeney felt when he walked into that conference hall for the first time. Duchenne is a devastating condition that makes itself known with a cruel subtlety that ensures maximum heartbreak. For a time, parents are afforded the bliss of watching their children develop normally; most will even see their child take those first joyful steps.

But gradually, they begin to notice that something is not quite right. Between the ages of two and seven, when a diagnosis is usually made, most children with Duchenne move slower and with more difficultly than other children their age. They might appear clumsy and fall, and have difficulty climbing, jumping, or running. They are often tired and always want to be carried.

Even so, "it can be difficult for parents to accept or believe the initial diagnosis," warns the website EndDuchenne.org. At times the children may appear to be improving, even as their muscles inside, unseen, are slowly ripping themselves apart.

The ambiguity fades in the second stage of the disease. Between ages six and nine, the child develops an eccentric walk to compensate for weakness in the trunk and thighs, sticking out the belly or throwing back the shoulders, walking on the toes or the balls of the feet. By twelve, many children need wheelchairs. Breathing and heart problems strike around age fifteen. The average life expectancy is just twenty-five.

At the conference, Sweeney explained how he believed an inability to make a single protein could wreak so much devastation and suffering—how it was that muscle wasting could arise from the equivalent of a single microscopic typo in the molecular instruction manual carried in the nucleus of every cell of the body.

Every human being has somewhere in the neighborhood of 20,000 different genes located in tightly coiled double helixes present in each cell's nucleus. Each of these genes comprises anywhere from 27,000 to 2.4 million pairs of DNA's core building blocks, microscopic clusters of molecules known as "nucleotides." Each nucleotide contains one of four key molecules known as "bases": adenine, cytosine, guanine, and thymine. The sequences of these bases, referred to by their first letters—*A, C, G,* and *T*—encode the molecular-level instructions our cells use to build every single protein our bodies produce. Those proteins in turn help determine everything from our hair color to our temperament to the ratio of our fast- to slow-twitch muscle fibers. It is a mistake in the sequence coding for the protein dystrophin that causes Duchenne muscular dystrophy.

Dystrophin is an exceptionally large gene that Sweeney compares to a "very stiff spring." It's a cellular shock absorber, necessary because actin and myosin are surrounded by a fragile cellular membrane. Dystrophin connects to this fragile membrane, joining these fibers to an elastic matrix on the outside and cushioning the force of the contractions in such a way that the membrane is protected. If the cells inside the membrane pull too hard, dystrophin yields, like a pliant spring absorbing the force and preventing the fragile cell walls from getting ripped apart.

Without this key cellular shock absorber, every time children with Duchenne move, they damage their muscles. (Imagine driving a car with no shocks over a road full of potholes.) Slowly the muscles begin to tear themselves apart. That's why children begin to acquire that awkward gait. Why they lose strength over time. Why—even as they grow weak—their muscles appear to bulge bigger than ever, offering false hope. The bulge doesn't come from added fibers made of myosin and actin. It comes from the buildup of fat and thick, unyielding scar tissue, hard clumps that will eventually condemn to a wheelchair the children forced to carry them.

At the end of Sweeney's talk at his first Duchenne conference, he found himself surrounded by the parents of those afflicted by the disease. Their tone was decidedly different from that of the small collegial crowds of students who would approach Sweeney after his lectures at Penn.

"These parents were just desperate," he recalls. "They were desperate to learn anything that they could that would give them some feeling, some understanding, about what was happening to their child."

Most of all, Sweeney remembers their confusion at the seeming indifference of the world, the feeling that they were alone and forgotten.

"They wanted to know why more scientists didn't want to work on trying to fix it," he recalls.

Suddenly Sweeney's deep intellectual curiosity, his delight in nature's secrets, turned into something more. Sweeney was abruptly thrust into a vortex of real human suffering, which altered the course of his career. "I felt guilty telling them that I didn't really work on trying to fix it, that I just worked on trying to understand it," Sweeney says.

When Sweeney returned home, he couldn't stop thinking about those parents and their children. He wanted to do something. And now his understanding of the mechanics of the disease took on a new color, a color of visceral tragedy.

If the cause of all this suffering and misery really was a mutation, the obvious solution was to try to find a way to reverse it. But how would he even begin?

The idea that we might actually rewrite the body's own instruction manual, the possibility that we could dig deep into the biological blueprints of the human body and make transformative changes in our DNA, is unlike any previous scientific pursuit in history.

Some would say we are hacking God's code itself. Certainly we are messing with genetic sequences honed over billions of years of evolution. That's why scientists have long warned that, if we are to proceed, we must do so with caution. Tinkering with DNA can have unintended consequences. We could unleash new diseases, mutant species of animals. We could create Jurassic Park.

At the same time, it's always been clear that the potential to cure human suffering is too great *not* to continue down this risky path. Researchers and doctors have long argued that if we can master genetics, the promise for curing disease will be almost unlimited. We can cure those children with Duchenne—and countless other infirmities. We can save lives. This is something scientists recognized almost as soon as DNA was discovered, despite concerns over the misuse of the technologies and the possibility they might benefit the few rather than the many.

Still, it has taken decades for that vision to reach the clinic. The quest began in earnest about forty years ago. In the late 1960s and early 1970s, researchers at John Hopkins University demonstrated for the first time that enzymes could be directed, like a pair of magical, microscopic scissors, to chop long strands of DNA into distinct fragments at any point. And soon after, biochemists at Stanford University published a series of papers describing how they had fused different strands of these fragments together by cleaving each one in such a way that the ends extruded complementary nucleotides. These nucleotides attracted one another like the opposite poles of a magnet. Scientists called the fused product "recombinant DNA."

In 1972 Theodore Friedmann and Richard Roblin articulated the radical implications of these techniques in a seminal paper in *Science* titled "Gene Therapy for Human Genetic Disease?" The future of medicine, they suggested, might lie in rewriting our own genetic blueprints.

In just the past several years, biologists took another huge leap forward, developing a new gene-editing technology, called CRISPR (clustered regularly interspaced short palindromic repeats), that is far easier, quicker, and cheaper to use than any technique previously available. Previous tools cost thousands of dollars and often took months to develop—changing just one gene could make for a student's entire thesis. Until relatively recently, targeted gene therapies worked by inserting genetic material into random locations on a chromosome, which sometimes caused unwanted side effects. The CRISPR technique, shown to work as a gene-editing tool on human cells in 2012, is far more accurate. It works by reappropriating a system used by single-cell organisms to keep track of foreign DNA from previously encountered viruses and plasmids that pose threats to the cell. Using "guide RNA" as molecular markers to precisely delineate the location of desired cuts in human cells, scientists have shown they can direct the actions of an enzyme known as "Cas9," which has the ability to slice open DNA, and then extract unwanted genes or insert new pieces of genetic material into a cell at will.

The technique is allowing technicians to perform the equivalent of microsurgery on genes, allowing them to precisely target and easily change DNA sequences at multiple locations across a chromosome. They can make these powerful modifications relatively quickly, using off-the-shelf technologies that cost as little as thirty dollars. Many believe it will soon allow the rewriting of complex diseases and traits caused by multiple genes.

But long before the advent of CRISPR, scientists were attempting to make use of modified DNA. In 1990, a team at the National Institutes of Health led by W. French Anderson treated a four-year-old girl with "bubble boy" disease by removing blood, isolating white blood cells in a petri dish, and then exposing them to a virus they hoped would inject its genetic cargo into her cellular nuclei.

The virus had been hollowed out and infused with recombinant DNA encoding the blueprints for the manufacture of a key enzyme needed to make infection-fighting T-cells—the very enzyme the girl's body had been unable to produce on its own. When the scientists returned the cells to the little girl's body and it started to crank out the enzyme she needed for the first time, it was a seminal moment in the history of science.

The effects Anderson and his team produced were only temporary and not as powerful as some hoped, because most of the little girl's old cells continued to contain the faulty DNA. As time went on, her diseased cells continued to divide at a rate far faster and more numerous than the small number of genetically modified cells Anderson had returned to her body.

It was one of Sweeney's colleagues and eventual collaborators at Penn, a biologist named James Wilson, who, two years after Anderson's pioneering 1990 effort with the bubble-boy patient, demonstrated a more permanent technique. He inserted a virus into the liver of a patient suffering from a genetic condition that created fatal levels of bad cholesterol. Since the liver is home to many regenerative cells, Wilson's technique was far more effective than any early trial—the modified liver cells multiplied quickly and in large numbers, and over time the organ turned into a steady source of new cells, becoming a production plant for missing enzymes and releasing them continuously into the bloodstream.

Later Wilson would almost see his career destroyed by another obstacle—the body's own infection-fighting machinery, which can sometimes respond to the presence of the viral vectors used to deliver the new DNA in unexpectedly violent ways. In 1999, Jesse Gelsinger, an idealistic, eighteen-year-old Arizona resident with a relatively mild form of a genetic disease, volunteered for one of Wilson's studies. Within a couple days of being injected with a virus carrying the modified DNA, Gelsinger's temperature had climbed to 104.5 degrees, and his body was overwhelmed with inflammation, indicating an extreme immune response. Finally, four days later Wilson's phone rang at four in the morning. It was the physician in the intensive care unit, informing Wilson that Gelsinger had been put on a cardiac bypass machine. His organs had begun to fail. Soon he was dead.

"The proteins that delivered the genes activated his immune system in a way we hadn't seen before," Wilson says. "It hit completely out of the blue. Each time the phone rang it was getting worse."

The tragedy led to lawsuits and congressional hearings, almost wrecked Wilson's career, and helped set the field of genetic engineering back years. Finding a way to suppress the sometimes virulent attack on the modified viral vectors used to deliver life-treating DNA would stand as the thorniest problem in gene therapy for much of the first decade of the 2000s—though in recent years researchers have made significant strides in overcoming it.

But there was and remains a larger, perhaps even more daunting obstacle faced by researchers hoping to advance gene therapy: the complexity of the genetic code itself.

The human genome is dizzyingly intricate. Unlike bubble-boy disease, and unlike Duchenne disease, the vast majority of human traits and diseases are caused by the interaction of many different pieces of DNA and environmental traits. And while scientists have been able to dabble in genetic engineering by targeting simple diseases caused by single mutations, providing evidence that the vision for genetic therapies that Friedmann and Roblin articulated is actually possible, and while CRISPR raises the possibility that even more complex diseases might be cured with targeted tweaks at multiple locations, in many ways the work has just begun.

Scientists are still struggling to decipher how it is that the constituent parts of the human genome and the environment interact to change us for better or for worse—indeed, scientists have only recently developed the very tools needed to quickly and cheaply read the 3.2 billion nucleotides that form the genetic sequence of a single human being. Most scientists recognize that only when these tools are refined will the full promise of gene therapy bloom.

Here is where the same advances in processing power, mathematical wizardry, and pattern recognition analytics that Hugh Herr is using in his lab to decode the workings of the human leg are also beginning to transform molecular biology. It took thousands of scientists more than a decade and somewhere in the neighborhood of $3 billion to decode and spell out the first 3.2-billion-nucleotide human genome, in 2000. Today companies can do it in a little more than a day, at a cost of well under $5,000 a pop. By the time you read this, they will likely be able to do it for $1,000—or far less. Every month, it seems, DNA "sequencing machines" grow more efficient, their price falls, and the possibilities of genetic manipulation grow greater.

To sequence genomes, scientists today are harnessing relatively new automated techniques that snip pieces of DNA into manageable fragments, rapidly make millions of copies, and then use sophisticated molecular-tagging techniques and visual recognition software to "read" the letters of individual genomes. These technologies, along with massive computing power to analyze and compare a growing library of fully sequenced, 3-billion-nucleotide genomes, promise to revolutionize our understanding of how different combinations of genes interact to cause disease and determine how we look, act, and think.

To see the front lines of this part of the revolution in action—and to see where the work of single-mutation genetic engineering like that done by Sweeney might one day lead—in 2014 I traveled to the humid, fog-enshrouded southern Chinese metropolis of Shenzhen to visit a company called BGI, formally known as the Beijing Genomics Institute. BGI is located in an eight-story former shoe factory near the port. Since 2008, it has grown from 20 employees to more than 5,000 and transformed into the world's largest single sequencing entity in the growing field of genetic research.

It's here and other places like it that researchers are characterizing and analyzing the vast sea of data, looking for the patterns that might explain how it is that those billions of microscopic proteins we all carry around in each cell interact to make us who we are, and how small changes can combine to cause our bodies to break down.

Today BGI's scientists and their collaborators are pushing the limits of what is possible using the tools of modern-day genetics. In addition to helping to unlock the genetic sequence of countless human disorders, they have identified the genes that can be tweaked or bred to create unusually large fish and high-yielding, drought-resistant millet. They have genetically engineered a new breed of miniature pigs that glow in the dark when hit with UV light—useful for scientific research, because it's easy to track transplanted organs. BGI's engineers have helped sequence the DNA of a four-thousand-year-old iceman found in Greenland, to find out how he differs from modern humans.

BGI, as it happens, has also ventured into that ethically murky territory that encompasses the search for genes that can make all of us bigger, faster, stronger, and smarter. They have, in other words, begun to search for the same kind of molecular-level typos that underlie those muscle mice and allowed baby Liam Hoekstra to grab his mother's fingers and pull himself up into the Iron Cross.

Recently, BGI began sequencing the DNA of more than two thousand individuals with exceptionally high IQs. The goal is to unearth the genetic basis of intelligence, a project the company has agreed to undertake in collaboration with researchers at King's College London and the University of Washington. It's no simple task: As many as ten thousand different genes (about half of those in the human genome) are believed to contribute something to individual intelligence.

During my visit to BGI's labs, I met Chris Chang, an American programmer and geneticist who was working on the project, and asked him about the project's goals and the controversy it has evoked in some quarters, from medical ethicists and researchers who worry it will result in genetically modified babies. "I think a world in which everyone is able to have smarter children if they want to is going to lead to a better society overall," he told me.

Many scientists are skeptical BGI will succeed in unraveling the mysteries of intelligence. But if BGI is successful in their search, it would not be the first time the company has unearthed genetic traits that might prove relevant for bioengineered human augmentation. After sequencing the genomes of fifty Tibetans and forty Han Chinese in 2010, the company announced they and researchers from the University of California had discovered more than thirty genes with DNA mutations that give some individuals an advantage at high altitudes. Nearly half were related to the way the body uses oxygen—in other words, potential levers for medicines or genetic manipulation that might facilitate artificial altitude adjustments.

The cautionary words of university ethicists worried about the implications of the coming genetic revolution seemed distant indeed as I roamed BGI's halls. In 2010, BGI inked a deal for a transformative $1.5 billion loan from the China Development Bank, which finances projects that are congruent with government policies. That allowed BGI to balloon virtually overnight from a relatively small operation to a company with far more genetic firepower than any other single research entity on the planet.

During my visit to Shenzhen and to a Hong Kong outpost located just across the border in a former printing plant, I passed room after room full of baby-faced Chinese technicians in blue lab coats, the best and brightest, leaning over vials with pipettes as they prepared samples for the sequencing machines.

Other researchers waited to carry these samples up flights of stairs to a cavernous, fluorescent-lit room lined with tables and spanning an area that extended at least half the length of a football field. I stepped into one of these rooms in Hong Kong. It hummed with the constant sound of powerful air-conditioning fans precisely adjusted to keep the temperature at 20 degrees Celsius (68 degrees Fahrenheit). On the ceiling, dark orbs—at least sixty of them—protruded every several feet, encasing cameras that stream images to a distant control room somewhere in the factory. They were pointed at the objects on the tables in front of me.

On each table sat a sleek rectangular device, slightly bigger than a mini-fridge but worth more than many a four-bedroom house in a nice American neighborhood. In 2010, BGI bought 128 of these $750,000 state-of-the-art genetic sequencing machines known as Illumina HiSeq 2000 Sequencers.

Each thirteen-day "run" by an Illumina machine produces 600 gigabases of information (600 billion nucleotides), enough genetic data to fill *six library floors,* each with a thousand yards' worth of shelving holding academic journals (or 1,200 times the amount of data a single CD-ROM can hold). Those six library floors would spell out the complete 3 billion nucleotide genomes of 200 individuals (or as BGI does it, 10 individual genomes sequenced 20 times for airtight statistical accuracy—which comes out to about 1,730 genomes every thirteen days). Somewhere within all this data lie patterns that might hold the key to much of what makes us who we are—and what molecular levers we might pull to allow those who wish to transform into whoever it is they might want to be.

After the Illumina machines spit out their data, it's the army of young workers across the border in Shenzhen, sitting in cubicles in a warehouse-like room, that will take the next step—cleaning up the data, and searching for correlations between specific letters in the DNA sequence and individuals who share characteristics or conditions that might eventually be traced back to specific genes.

To manage and compare all this, BGI has constructed several supercomputer centers. Analyzing the DNA is a mathematical challenge exponentially more complex than teasing out the relationships between the parts of a human arm or leg—a challenge, as we learned in the last chapter, that even in the realm of biomechanics was so far beyond previous generations of engineers that it's only now that people like Hugh Herr and Patrick van der Smagt are able to model the interactions between thickets of variables and tease out their relation to one another and to movement.

With so many more variables, billions to Herr's thousands, BGI's appetite for processing power is never satiated. And the amount of memory at the company's disposal is constantly growing. The goal is 1,000 teraflops of capacity (or 1,000 trillion operations per second); when I visited, the company had recently announced it had surpassed one-quarter of that. For individual projects that require BGI's statisticians to run sophisticated regression analysis on multiple 3-billion-nucleotide sequences at once, the Chinese government allows BGI access to several of the largest supercomputers in the world, located at government-owned-and-run computing centers.

BGI's greatest discoveries are likely yet to come. The company has announced plans to sequence 1 million human genomes, an ambitious goal that would give them a genetic library unlike any ever before created. The company hopes, in other words, to read 3 quadrillion nucleotides, producing enough data to fill *30 million library floors*. Such a vast trove of data would presumably allow researchers to search at will for correlations between specific genes and specific traits or diseases and yield powerful statistical certainty. It would also, of course, require an unprecedented amount of computing power to master.

I asked Chang, the intelligence project researcher, if the prospect of any engineered human traits made him uncomfortable. I repeated to him a fear often invoked by medical ethicists—that the technologies will benefit a select few and leave the rest of us in the dust. Chang did not seem particularly concerned. But when pressed, he admitted that some scenarios did gave him pause.

"If you have direct gene-editing capability, then some of the things you can imagine are kind of scary," he acknowledged nonchalantly.

Like what? I asked. After all, many believe CRISPR will give us direct genome-editing capability.

Chang singled out the traits associated with sociopaths like Bernard Madoff and Charles Manson, who seem to have felt no remorse for committing acts that most of society would find repugnant.

"If someone feels that it would give their child an advantage," he explained, "and they know how to engineer . . . just the right combination of a lack of empathy and extreme confidence, then that's a scary thought."

The more Lee Sweeney thought about helping those desperate parents and their children with Duchenne disease, the more he realized he also wanted to find a way to help a second group of patients suffering from the scourge of muscle wasting. As it happened, at the time of that transformative Duchenne conference, Sweeney had already been ruminating on the ravages of aging. For months he'd been both grief-stricken and haunted by the all but inevitable muscle deterioration that turns the oldest among us into frail shadows of our younger selves.

It was the death of his grandmother, Mattie Theo Richardson, that had started him thinking. For years Richardson had been living happily with Sweeney's parents in Arlington, Texas. But the end, which came at age ninety-one, was not pretty. Richardson had always been an energetic woman who delighted in getting her hands dirty in the garden. With age, though, she'd grown increasingly weak, until one day her legs failed her. Richardson fell and broke her hip. She lived for another year and a half, but she never recovered.

The last time Sweeney saw her, Richardson told him she could no longer do the things she loved doing, that she was too frail and there was no point in living anymore.

"From that, she basically wasted away," Sweeney says. "Because of her muscles being so weak, she allowed herself to die."

The death had prompted Sweeney, in the months before he accepted the invitation to speak at that Duchenne conference, to take a closer look at what happens to our muscles as we age. Between the ages of thirty and eighty, all of us lose on average about one-third of our skeletal muscle mass. We all literally begin to waste away. Sweeney wondered why—and if—it had to be that way. It seemed to him that the same raw materials used to build muscle in the young remained available in the bodies of the old. What was it that caused the body to suddenly stop repairing muscle and building new muscles as needed?

In the stories of suffering he heard from the parents of children with Duchenne disease, Sweeney heard echoes of the muscle wasting he'd become preoccupied with in aging. In the helplessness of those desperate parents, Sweeney recognized his own. Sweeney realized that if he could crack the mystery of age-related muscle wasting, it could also have significant benefits for those with Duchenne. Giving patients with Duchenne bigger muscles as he'd dreamed of doing for his frail, bedridden grandmother would also buy these children and their loved ones precious time and improve the quality of that time.

There was another reason to try. Sweeney had been closely conferring with the geneticist Jim Wilson and his colleagues. The two had even published a paper together on gene therapy and dystrophin. It was true that Duchenne was caused by mutations affecting this single protein, but the dystrophin gene is the largest gene found in nature. It consists of at least eight independent, tissue-specific promoters, and measures about 2.4 million nucleotides. The protein itself contains more than 3,500 amino acids. Those viruses scientists had found a way to hollow out and transform into delivery mechanisms for man-made genetic materials simply weren't large enough to hold the instructions for dystrophin. The DNA didn't fit.

So as Sweeney and Wilson worked on the dystrophin problem, Sweeney resolved to try to do something immediately—something that would help both those with Duchenne and victims of age-related wasting, like his grandmother, Mattie Theo Richardson.

Sweeney started by attempting to diagnose the problem in the elderly. He wasn't entirely sure why we lose muscle as we age, but he suspected the answer might lie in the age-related slowdown of the endocrine system, a group of glands that broadcast global instructions across the body—from initiating our fight-or-flight instinct, to signaling the body that it's time to go to bed, or telling us when we're in love—by releasing hormones into the bloodstream.

Sweeney knew that hormones of the endocrine system also played a role in triggering and regulating muscle growth and repair. In fact, both muscle-building synthetic steroids and genetically engineered human growth hormone (HGH) work by mimicking compounds released by the body's endocrine system—hormones released at levels that, Sweeney knew, plummet in our body as we grow older. If Sweeney could somehow target the areas of the muscle that were taking their cues from these global growth signals—if he could find a way to send them a message on his own—perhaps he could convince them to grow. Sweeney decided to hack the system.

He considered his options. Anabolic steroids were out of the question. Both the elderly and those with Duchenne often had cardiac problems, and a growing body of research suggests that anabolic steroids can weaken both the heart's ability to effectively pump blood and its ability to relax and refill with blood between contractions. Treating them with steroids might build skeletal muscle, but if he made the heart worse, the patients wouldn't be around long to enjoy it. Anabolic steroids, Sweeney also knew, were a long way removed from the molecular switches he hoped to access. They are essentially modified versions of testosterone, the male sex hormone (remember those testicle-chewing Greek Olympians from the introduction?); they also cause facial hair and other symptoms entirely unrelated to muscle growth. Sweeney suspected that changes in a different part of the endocrine system caused muscle wasting in the elderly—after all, it hit both sexes.

Next he considered HGH. HGH is released by a pea-shaped structure at the base of the brain called the "pituitary gland," which sends out a signal to build body mass (and is thus also a popular target for doping athletes). And indeed, the idea that changes in the pituitary gland might underlie the changes in an aging body made a lot of sense.

But though HGH seemed a promising target, to Sweeney it also seemed a little too far removed from the mechanism of muscle growth he hoped to access. Hormones work by a "lock and key mechanism." They circulate through the bloodstream until they come across proteins that they fit into. The proteins, known as "receptors," poke out from different cells all around the body. When the hormone binds to a receptor, it starts up processes in the cell's DNA, in the same way that sticking a key in an ignition switch starts up a car.

One of the ways both testosterone and HGH were believed to build muscle mass was by signaling the body to produce a third compound, called "IGF-1." It's IGF-1, produced by the muscle cells themselves, that actually sets in motion a cascade of chemical processes that lead to more growth. Once this happens, stem cells, the body's itinerant construction crews, get to work in the muscle cells in our bodies, adding new layers of myosin and actin, those two filaments that slide against one another and allow our muscles to contract, converting chemical energy into the kinetic oomph that allows us to move and act in the world.

"In the end we decided we'd go straight to the IGF-1 because that is what we really wanted anyway," Sweeney says.

In other words, he decided he would attempt to turn on the IGF-1 gene itself (rather than relying on hormones to do it for him)—and send out the equivalent of a work mobilization order to scores of tiny protein worker bees within the muscle cells themselves that were capable of construction and repair. But instead of a dose of hormones that would eventually dissipate, he would use an engineered virus to deliver an artificial gene directly to the muscle that would remain switched on. Permanently.

Sweeney created three pools of test mice, at the mouse equivalent of what in humans we consider young (two months), middle-aged (eighteen months), and elderly (twenty-four months). Then he injected the virus into the muscles of each right hind limb. He left the left limbs alone to serve as comparisons. Four to nine months later, he sacrificed the mice, cut them open, and examined their muscle growth.

The results were unequivocal. The muscle mass of the youngest mice increased by 15 percent and their muscle strength increased by 14 percent. But it was when Sweeney examined the older mice that he was most impressed. Just by eyeing the results, it was obvious to Sweeney from the start that there was more muscle there—that the IGF-1 had powerfully jump-started the process. But when Sweeney walked into the lab one day, parked himself in front of a computer screen, and began to go over the data with a postdoctoral researcher, he was shocked by the numbers.

Sweeney had expected some muscle improvement. But when he analyzed the muscle of mice who, in mouse years, were the equivalent of a ninety-year-old human—the very age at which his beloved grandmother had succumbed to her weakness—Sweeney found that their muscles were just as strong and healthy as the muscles in the young animals.

"Well, that's a little better than we expected," he declared, in his understated way.

"We expected that they would be better," Sweeney now says. "But not that they would be as good as they ever had been in the animal's life."

In the older animals, the muscles grew by 19 percent, while strength increased by as much as 27 percent. Furthermore, the IGF expression "completely prevented the significant loss of the fastest, most powerful fiber types," implying not just that new muscle production was increasing, but that muscle regeneration was amped up as well. The regeneration suggested that the technique might help patients afflicted with muscular dystrophy by helping to preserve muscle function. For the rest of us, the implications were equally profound.

"When the animals got super-old their muscles never changed," he says, calling to mind images of demented ninety-year-olds competing at the CrossFit Games. "They stayed just as strong as when the animals were young."

When Sweeney published the results in a scientific journal, he included a postscript in the paper that was unusually personal for a dry scientific journal. It read: "H.L.S. dedicates this work to the memory of his grandmother, Mattie Theo Richardson, whose life was diminished and ultimately shortened by lack of sufficient muscle strength to stand and walk."

Even though Sweeney's paper had deep personal meaning to him, he also recognized the danger. For athletes, he warned when he released the paper, IGF-1 gene therapy could be the "perfect performance enhancer."

"You build muscle mass and strength even without exercise," Sweeney said. "And it is not detectable in the blood." The technique, he quipped, "is a couch potato's dream."

Still, Sweeney was unprepared for the magnitude of the response—that flood of calls from desperate athletes.

In 2001, a pair of reporters from the British newspaper the *Guardian* visited Sweeney's lab and viewed a member of the newest generation of his mice up close. When Sweeney trotted out the comically ripped rodent, the reporters gave it the nickname "He-Man."

By targeting the liver, Sweeney had created a centralized production plant for amped-up IGF-1 molecules, which then circulated through the bloodstream throughout the body, leading to global muscle production. The treatment increased He-Man's muscle mass by a breathtaking 60 percent, allowing him to "effortlessly" scale a ladder with 120 grams of weight—three times his body weight—strapped to his back.

By this time it had been about three years since Sweeney first introduced the mice to the world, and his phones had been ringing off the hook with calls from athletes ever since. The experience had begun to color his perception.

"I honestly believe that if the Soviet Union hadn't fallen apart, it would be genetically altering humans by now," Sweeney told the reporters at the time. "Who knows where it would go?"

Many other researchers share similar concerns. This concern extends well beyond those just working on muscles—as I found out firsthand. In his book *Run, Swim, Throw, Cheat,* the British sports scientist Chris Cooper writes that sporting achievement requires "power, endurance and the ability to compete beyond the pain barrier." Pro athletes intent on cheating have attempted to augment these qualities using what Cooper calls the "unholy trinity"—anabolic steroids, the hormone erythropoietin (EPO), and stimulants—for decades. And indeed, one of the more colorful and improbable gene-hunting stories I came across involved not the pursuit of a muscle mutation, but that third component of Cooper's "unholy trinity," pain tolerance.

In the early 2000s, a team of British geneticists received an unusual report from doctors at a hospital in Lahore, Pakistan. The doctors spoke of a ten-year-old Pakistani street performer who made his living by sticking knives into his arms and walking across hot coals. The boy's wounds from these acts were very real—he continually showed up at the hospital after performances bleeding and burned, asking to get patched up. But here's the strange part: No matter how bad the wounds of the day, the child performer seemed not to feel the slightest hint of discomfort. He seemed, in fact, barely to mind at all. The doctors began to suspect he had a rare mutation that made him immune to pain.

Sadly, by the time the team arrived to test this theory, it was too late. The boy had leapt off a roof to impress his friends, and died. The geneticists, however, *were* able to collect DNA samples from the street performer's clan. There they discovered that individuals in three families had a defect in a gene known as "sodium channel N9A (SCN9A)," one of eleven human genes coding for proteins involved in the initiation of pain signals sent through the body. These individuals could have all walked across hot coals if they'd wanted to, and it wouldn't have hurt at all. The mutation is a promising pathway for painkilling drugs. But one can imagine why the ability to run through pain would prove just as powerful in some sports as bigger muscles are in weight lifting. Indeed, it raises the disturbing possibility of not just athletes but of soldiers being willing to do what it takes to win or of authoritarian governments forcibly changing soldiers into pain-free cannon fodder.

The study's senior author, C. Geoffrey Woods, of the Cambridge Institute for Medical Research, was reluctant to talk about the work, hanging up when I contacted him. He was not the only one in the field who was cautious. Se-Jin Lee, a professor at Johns Hopkins University School of Medicine, is, like Sweeney, considered one of the world's foremost experts on muscle-signaling pathways. But Lee ducked my calls. He is reticent with journalists, he explained to David Epstein, author of the 2013 book *The Sports Gene,* because "he is troubled by the apparent willingness of athletes to abuse technology that isn't even technology yet, and that is meant for patients with no other options." Lee worries that the kind of treatments he and Sweeney work on involving muscles might get stigmatized the way steroids have "because of their role in sports scandals."

Certainly steroids provide a cautionary tale. It's easy to forget that long before they became associated with cheating, anabolic steroids were used as a powerful force for good to treat the weakest among us—ranging from the malnourished, skeletal concentration camp survivors liberated from places like Auschwitz at the end of World War II, to burn victims, to children suffering from growth problems.

For a textbook example of why that changed—a cautionary tale very much on the mind of Sweeney, Lee, and their peers as they contemplate a new era of gene doping—it's worth considering the experience of a man named Gordon Hughes.

In 1961, Hughes invented a new compound called "norbolethone" or "Genabol" as part of his Ph.D. chemistry thesis at the University of Manchester, in England, believing it might help elderly surgery patients who needed to build more protein. Wyeth Pharmaceuticals (now part of Pfizer), where Hughes eventually worked, *did* research the compound for use as a weight gain treatment for those with short stature.

But most of the world didn't learn about the drug in the context of elderly surgery patients or children of short stature. I certainly didn't and neither did Sweeney. I learned about the drug a few years ago when I flew out to the cornfields of Champaign, Illinois, to meet Patrick Arnold, a gruff, barrel-chested chemist who discovered Hughes's compound while researching steroids in a medical library. Arnold had a very different use in mind. In the early 1990s, Arnold was a twenty-four-year-old old lab technician working a low-paying job that he hated, making the chemicals used in shampoos and hair gels. Arnold was also an avid weight lifter and he was bored.

One afternoon after starting the day's reactions at work, Arnold looked up the molecular structures of the steroids mentioned in his muscle magazines, and had a revelation: "I hate my job, I'm sitting here, I've got a lab—I can try making some of these things myself," Arnold recalls thinking. "No one will even know what the hell I'm doing."

Arnold added the steroid precursors he would need to the regular list of laboratory chemicals he ordered through the company, and nobody noticed. Soon he was spending ten hours a week visiting libraries, combing through obscure patents and research journals for compounds with molecular structures worthy of further exploration. One of the most exciting to Arnold was Hughes' norbolethone. It had a unique chemical structure that would be impossible to detect, but it also seemed to have many characteristics of the more potent steroids Arnold had tried.

Years later, when Arnold had become a successful mainstream nutritional supplements guru whose products were used by professional baseball players like Mark McGwire, Arnold brewed up a batch of norbolethone just for fun. Norbolethone was so obscure that professional doping programs had no reference sample and thus no way to test for it. But then Arnold passed the compound on to Olympic cyclist Tammy Thomas. Completely ignoring Arnold's caution on dosing, Thomas took enough of the compound that she could soon squat-lift an impossible 315 pounds. She also developed a prominent Adam's apple, a deep male voice, facial hair, and male-pattern baldness. Eventually her natural testosterone levels (females produce some, too) dropped to levels so far below normal it set off alarm bells. (This is a common side effect that occurs after a course of steroids is completed.) Once testers began to scrutinize her urine, it was only a matter of time before they identified metabolites that led them to norbolethone.

Arnold had also sent a sample to a brash entrepreneur named Victor Conte. Conte ran a sports nutrition center in Burlingame, California, called the Bay Area Laboratory Co-operative (BALCO). Conte rechristened the steroid "the clear" and began distributing it to top athletes. After Thomas got busted, Arnold did something almost unprecedented in the annals of sports doping. He went through the Merck index of compounds and created an entirely new steroid.*

Eventually, BALCO's activities would blow up into a major scandal that would ensnare some of the nation's most prominent athletes, tarnishing the names of baseball greats Jason Giambi and Barry Bonds, football star Bill Romanowski, British sprinter Dwain Chambers, and three-time Olympic gold medalist Marion Jones, among others. Arnold was sentenced to three months in federal prison in Morgantown, West Virginia.

It's only because of Arnold's famous concoctions that most people have heard of the work of Gordon Hughes. The retired scientist does not appreciate the notoriety.

"It bothers me that people are getting these drugs," Hughes told me. "They haven't passed FDA requirements. You don't know anything about them."

Back when Hughes first made the compound, he said, it never occurred to him that his creation could end up being used that way.

But it has occurred to officials at the World Anti-Doping Agency (WADA) as they contemplate the coming era of gene therapy. In fact, BALCO is very much on their minds. Every January, Sweeney, along with a number of other expert geneticists, makes his way to Montreal and heads to a gleaming, forty-eight-story skyscraper on the city's two-hundred-year-old Victoria Square. There he spends the following eight hours cloistered in a conference room on the seventeenth floor, with panoramic views of the city, eating whatever food is brought in and discussing the myriad ways in which athletes might abuse the emerging science of genetic engineering to alter their bodies.

"They're [WADA] hoping to be ahead of things this time," Sweeney says. "They're hoping the athletes will be more afraid to

---

* Eventually that too would unravel when a track coach fished a used syringe out of the garbage at a competition and sent it to a doping watchdog agency.

go after it and they won't be caught off guard like they were with the whole BALCO fiasco."

Still, WADA is limited in what it can do. One possibility is to contact companies doing gene therapy trials, obtain samples, and search for any biological signature that might linger in the body and reveal that an athlete has tried to augment his or her genes. Another is to dole out grants to researchers trying to come up with innovative ways to test for gene doping.

The problem is that the only way to prove an athlete has hacked his or her genes is to detect the presence of the vector that delivered the new DNA. But the body will eventually break down the vector and remove any trace. Once that happens, it's virtually impossible to prove the DNA is not God-given.

"That is one of the things we discuss," Sweeney says. "You'd have to think about what periodicity of testing you'd have to have to make sure you catch them."

Perhaps the most daunting obstacle is the sheer number of possibilities, because tweaking the kinds of genes Sweeney has successfully tweaked in mice—and, as of 2011, golden retrievers—is only one possibility. WADA, Sweeney says, is "concerned about just about everything you could possibly think of that would give an athlete an advantage that you could do with gene therapy."

In the area of movement, at last count, there were more than *200 genes* that had been identified that are associated with superior athletic performance—entire books have been written about them.*

* Perhaps the most widely cited, single-beneficial athletic mutation augments endurance. Researchers identified the mutation by analyzing the DNA of Eero Mantyranta, a Finnish cross-country skier who won gold medals in the 1960 and 1964 Winter Olympics. Mantyranta, who died in 2013, had abnormally high levels of the protein hemoglobin in the red blood cells. Hemoglobin gives the blood its red color, and its job is to carry oxygen molecules to Type I muscles, where they can then use the oxygen to make the ATP the muscles need as energy to fuel contractions. The more he-

(Epstein's *The Sports Gene* is a good one.) Though at this point, the vast majority of these genes on their own play too small a role in isolation to be worth tweaking using the current technologies, gene-editing technologies like CRISPR are advancing quickly, suggesting that that might not always be the case. Meanwhile, a number of artificial genes are already within reach, like the pain gene, and IGF-1, and another mutation that would soon capture Sweeney's attention—the same mutation we learned about in the beginning of this chapter, in a German newborn. He was born with muscles so ripped the press called him "Super Baby." That mutation involved the obscure but powerful compound myostatin.

Myostatin is the reason I wanted to speak with that John Hopkins molecular biologist who wouldn't come to the phone, Se-Jin Lee.

---

moglobin you have, the more oxygen you can deliver to muscle fibers, the more energy you can burn, and the longer you can run without tiring. It's a benefit in endurance sports and exactly the effect many athletes artificially seek through the hormone erythropoietin (EOP).

Another mutation that appears to improve endurance seems to increase the production of tiny molecular organelles known as "mitochondria." By targeting a gene that "upregulated" mitochondria production, a team that included scientists in Korea and San Diego's Salk Institute managed to create a new breed of "marathon mice"—mice capable of running up to twice the distance of their normal, unmodified kin, without tiring. Mitochondria are tiny free-floating structures outside the nucleus of muscle cells; scientists often refer to them as the "power plants" of the cell. It is the mitochondria that use oxygen to convert sugars to the ATP that powers muscle contractions in slow-twitch fibers—and that can do so almost indefinitely, in a process that is called "aerobic respiration." (Type I muscles use a different pathway to produce "anaerobic respiration.") In 2007, researchers at Harvard managed to convert mice into endurance runners by targeting yet another gene that caused them to produce more of a little-known type of muscle fiber caked IIX fibers, which scientists have only recently begun to understand.

Lee discovered this new protein, which appeared only in muscle, in the 1990s, right around the time Sweeney was beginning his own study into dystrophin.*

Myostatin's role in the body, Lee and his graduate students discovered, is to inhibit muscle growth. If IGF-1, the compound targeted by Sweeney, is a muscle growth accelerator, myostatin is the brake. Without it the muscle grows unchecked, often to at least twice its normal size. Yet mutations that seemed to "knock out" the genes in animals, it turns out, are found in nature. Soon after Lee and his team produced myostatin-deficient rodents so large the media dubbed them "mighty mice," the team identified a similar, naturally occurring mutation in a breed of hypermuscular cattle called "Belgian Blue." Another team discovered a myostatin mutation in whippets, a breed of racing dog that can run 35 miles an hour. Dogs with two copies of the defective gene were muscle-bound and

---

* Lee had been looking for new members of a family of proteins known as "transforming growth factor beta superfamily" (TGF). Several related proteins in this family had already been identified that had surprising and powerful effects on, among other things, bone growth and cell division. Perhaps, Lee suggested, this new protein, which he called TGF-8 and would eventually be named "myostatin," might also have therapeutic use. To find out, Lee assigned a graduate student named Alexandra McPherron to find out what it did. McPherron bred mice to lack both copies of the gene that coded for the protein. McPherron noticed something odd almost immediately. The newborn animals "looked lumpy." It looked to her, in fact, as if all around the body, dispersed in a uniform way, the mouse had sprouted benign tumors, or oversize, oddly shaped, fur-covered warts.

It wasn't until McPherron euthanized one of these mice and dissected it that she discovered the true cause of the lumps obscured beneath the fur. The lumps were muscles. And the muscles were everywhere. The shoulders were huge. The abs were so thick, they were hard to cut with a scalpel. The arms and legs were straight out of a cartoon. McPherron was about to experience a eureka moment very much akin to the one Sweeney had experienced himself when he first gazed at those oversize mice legs in his lab. "When I opened it up, I kind of just sat there quietly thinking, am I really seeing this?" she recalls. "Then I started going to get other students to come over and look at it."

worthless on the track. But dogs with a single copy seemed to hit the sweet spot. They were often champions.

Within just a couple of years of Lee's discovery, doctors in a Berlin hospital contacted Lee. They believed they had identified the first human newborn with the mutation. It was the baby who would come to be known as "Super Baby."

"Normally if you pick up a baby, it feels soft because of the fat tissue that small children have around them," says Markus Schuelke, the child neurologist who examined the baby soon after his birth when the nurses noticed him experiencing tremors. "This child had a hard body. It felt more like muscle."

Schuelke had read Lee's paper. In 2004 the two researchers and several others published a paper in the *New England Journal of Medicine* reporting that they had verified the myostatin mutation in the baby. The resulting media frenzy upset the family. Some critics demanded the unnamed mother, whom the paper identified as a professional sprinter, return any medals she'd won. Today Schuelke will not release any information about the current condition of the boy—after that frenzy, the family asked the doctor never to release any information about them again. But the parents of perhaps the most widely known publicly identified "superbaby" in the United States do still speak with reporters.

Michigan's Liam Hoekstra, the boy we met at the beginning of the chapter, is nine years old at the time of this writing. He has a six-pack and a back that bulges with muscle. He is playing hockey and likes to wrestle, according to his father, Neil. And though he's not particularly competitive in the hockey rink, his strength, it seems, is giving him a distinct advantage in the wrestling ring, where he is able to overpower opponents, even without knowing traditional moves. Liam also can throw and hit a baseball much farther than other kids his age, his father says. And the muscles come in handy on the schoolyard. Liam, Neal reports with a hint of pride, recently "decked" a bigger kid who was bullying his friends.

Doctors still haven't pinned down the exact mutation causing Liam's hyperdeveloped muscles. But Sweeney, Ferrell, who examined his DNA, and others believe the cause is likely to be genetic. If that mutation is ever discovered it could point the way to new therapies—or new ways for muscle-heads and aspiring athletes to bulk up.

For now, to Sweeney and a number of others, myostatin is among the most promising targets to treat muscle-wasting diseases. And at the time of this writing, several companies have so-called myostatin inhibitors in the works.

In 2011, Sweeney used a myostatin mutation to create an Arnold Schwarzenegger golden retriever. In 2015, scientists in Chinese labs announced they had used CRISPR to create beagles with double the normal muscle mass by deleting myostatin, and intended to create dogs with other mutations that mimic human diseases such as Parkinson's and muscular dystrophy.

Sweeney is still interested in studying IGF-1, but he focused on myostatin inhibitors because they can be given systematically with few side effects, whereas IGF-1 cannot. He may yet return to IGF-1. After the 1999 death of Jesse Gelsinger, Wilson's institute was shuttered and he was banned from working on human clinical trials for five years. But since that time, doctors have found new ways to overcome the immune response, including the use of other vectors, some discovered by Wilson, and the application of certain steroids to hold inflammation in check in the crucial early phase of the therapy. The first genetic therapy was approved by the European Union in 2012. Today there are by some estimates more than two thousand genetic trials in process. The widespread use of these therapies to cure disease in the future seems all but assured. In the meantime, Sweeney and others are watching trials closely to see which viral vectors are safest.

On the long car ride back to New York City from Sweeney's office, I contemplate my own genetic makeup. I played sports when I was kid, and at one point dreamed of being a professional baseball player. I even tried high school football for a few months. Unfortunately, I was small for my age at fourteen. Thus one of my most vivid memories is of the afternoon I heroically placed myself directly in the path of a hulking, overdeveloped running back named John Burke after the quarterback gave him the ball. I was intent on bringing him down, come what may. As Burke steamrolled over me, I hugged his ankles like a helpless Lilliputian clinging to Gulliver. Burke dragged me at least thirty yards, probably without noticing, before depositing me in a demoralized heap somewhere north of the fifteen-yard line.

How different things might have been had myostatin inhibition existed back then . . . For a moment, I allow myself to imagine a world in which I might have the option of playing the role of John Burke, mowing over some other underdeveloped chump. Then it occurs to me that steroids were around back in high school. As was the weight room. And I never contemplated experimenting with either one of those.

Like Sweeney, I conclude, I can't relate to the idea of rewriting my genome in the service of athletic glory. But it might be different if the mutations more directly affected my livelihood—if, say, I could tweak my recall or intelligence. Indeed, as I would learn later—and as we will explore in upcoming chapters—some recent discoveries offer up this tantalizing and morally troubling possibility, suggesting we may soon be able to tweak our genes to improve memory, cognitive functioning, even happiness.

But on that ride home, it is the sad specter of those Duchenne patients and their desperate parents that sticks in my mind. The possibility of a few meatheads genetically modifying their muscles seems to me a small price to pay, a small risk to take. That is why Sweeney's work and that of researchers like Sweeney and Se-Jin Lee continues apace.

Gene therapy is not the only method of hacking into the machinery of the body and directing it to remake itself. In recent years, scientists have begun to explore other kinds of techniques with equally transformative potential. These techniques aim to unleash the latent powers of regeneration already present in the cells of our bodies—amazing powers that scientists have long suspected were there, but have only recently figured out how to access.

CHAPTER 3

THE MAN WITH THE PIXIE DUST

REGENERATIVE MEDICINE AND THE QUEST TO REGROW LIMBS

It was so gradual at first that he could easily have missed it: a faint pulse in what remained of his mangled right thigh muscle. Then the pulse became something more. Some people would have thought it impossible. But Corporal Isaias Hernandez could *feel* his quadriceps getting stronger. The muscle was growing back.

When he'd first arrived in the trauma unit of San Antonio's Brooke Army Medical Center in December 2004, Hernandez's leg had looked to him like something from KFC, "you know, like when you take a bite out of the drumstick down to the bone?"

He'd been hit while walking across a desert air base in western Iraq, clutching a purple and black twelve-inch television to his chest. The plastic TV shielded his vital organs from the artillery shell's shrapnel; his buddy carrying the DVDs wasn't so lucky. He didn't make it.

The doctors kept telling Hernandez he would be better off with an amputation. More mobility with a prostheses. Less pain. When he had refused, they'd taken a piece of muscle from his back and sewn it into the hole in his thigh. He'd done all he could to make it work. He'd grunted and sweated his way through the agony of physical therapy with the same red-faced determination that got him through boot camp—he'd even snuck out to the stairwell, something they'd said his body couldn't handle, and practically dragged himself up the steps until his leg had seized up and he'd collapsed.

But people didn't recover from wounds like his. The artillery round had ripped off 90 percent of his right thigh muscle, and he'd lost half his leg strength. Remove a large enough percentage of any muscle and you might as well lose the limb—the chances of regeneration are just as remote. The body kicks into survival mode, pastes it over with scar tissue, and leaves you to limp along for life.

For Hernandez, it had been *three* years and there was no mistaking it. He'd plateaued. Lately, the talk of amputation had cropped up again. The pain was constant and he was losing hope.

Then he'd seen a television program on Discovery Science and everything changed. The episode told the story of a Vietnam veteran in Cincinnati named Lee Spievack, whose fingertip had been severed by the propeller of a model airplane. Spievack's brother, a Boston surgeon, had sent him a vial of magic powder and told him to sprinkle it onto the wound. They called it "pixie dust." The fingertip had grown back.

That's when Hernandez remembered: Hadn't that doctor he met when he first got there mentioned some kind of experimental treatment where you could "fertilize" a wound and help it heal?

The doctor's name was Steven Wolf, and in February 2008, after Hernandez tracked him down, Wolf agreed to use the twenty-two-year-old marine as a guinea pig. First he put Hernandez through another grueling course of physical therapy to make sure he had indeed pushed any new muscle growth to the limit. Then Wolf cut open Hernandez's thigh and inserted a paper-thin slice of the same material used to make "pixie dust"—part of a pig's bladder, known as an extracellular matrix (ECM). Finally, he assigned the young soldier another course of punishing physical therapy.

Soon something remarkable began to happen. Muscle that most scientists would tell you was gone forever began to grow back. Hernandez's muscle strength increased by 30 percent, then 40 percent. It hit 80 percent after six months, went to 97 percent, and is now operating at 103 percent of what it was before the operation. Hernandez's muscle mass increased by 11 percent in those initial few months and has continued to grow ever since. Today he can do things he simply couldn't before, like ease himself gently into a chair, instead of dropping into it, kneel down, ride a bike—or climb stairs, without collapsing.

Two years later, a team of researchers based at the University of Pittsburgh's McGowan Institute for Regenerative Medicine won approval to begin an eighty-patient study at five institutions that will use this same ECM to regenerate the muscles of patients who have lost at least 40 percent of a particular muscle group—an amount so devastating to limb function that it often leads to amputation. Hernandez was the first patient to volunteer. He's hoping to regain even more strength and return to active duty.

If the trial succeeds, it could help fundamentally change the way we treat patients with catastrophic limb injuries, and begin a process that its promoters hope will someday "put the prosthetic industry out of business."

To some, Hugh Herr's adventures in bionics and Lee Sweeney's genetic engineering efforts sound like science fiction. But there's a third frontier in the augmentation of human movement that is in some ways even more fantastical. At leading universities around the nation, bioengineers are learning how to harness and enhance the capacities of the microscopic army of cells and signaling agents in charge of building and repairing different parts of our bodies. With that knowledge, they are attempting to push things much further than Sweeney and Lee, who in some ways have been only concerned with turning up or down the cellular volume switches already present in our muscles. The bioengineers in the field known as "regenerative medicine" are coaxing cells to do things that until only a few years ago seemed impossible—regenerating shattered bones and shredded muscles normally gone for good, growing new human organs outside the body and inserting them into patients, spraying skin cells out of aerosol cans onto burn victims.

Some have even made rudimentary progress in regrowing severed limbs—in the same way salamanders can regrow a tail. Others have transplanted the hands, feet, even faces of deceased donors onto patients. All of this hints at a potentially paradigm-shifting moment for the human race—the idea that someday in the not-too-distant future we may be able to regrow or swap in new parts as easily as we swap in a new set of tires on a car. It's a development that could redefine the experience of aging and vastly improve the quality of life for millions of people.

While this development itself is new, the idea has been around a long time. Researchers have wondered for centuries at the mysterious mechanisms that limit regeneration in some species and allow it to flourish in others. A salamander can grow back its legs, its arms—even its eyes. Lobsters can grow new claws. Some worms can grow new brains. Why is it, many wondered, that humans cannot do the same?

By the eighteenth century, these phenomena had been so well characterized in lower animals that the French philosopher, satirist, and author Voltaire, after cutting off the head of snail and watching it regrow its head, wrote a letter to a blind friend suggesting that humans would soon unravel these mysteries and do the same.

And indeed, throughout recorded history there has also been tantalizing evidence that perhaps this ability *does* lie dormant in our own bodies, waiting to be discovered, studied, and harnessed. Exhibit number one is the notorious variety of cancerous growth discovered, most famously, in Lance Armstrong, but referenced far back in antiquity, a bizarre kind of cancerous tumor called a "teratoma," the Greek name for "swollen monster." While most tumors contain only one type of cell, teratomas often grow into a jumble of many different types and tissues, mixed together in a large, frightening lump. They might contain pieces of bone, muscle fiber, nodules of cartilage, fluid, wads of hair, and baby teeth. They might pulsate with pieces of heart tissue, or quiver with fat. Relatively rare, and usually found in the ovaries or the testes, teratomas have occasionally appeared in the neck, heart, liver, stomach, spinal cord, and even under the eyebrow.

It's a terrifying prospect—imagine finding a tumor with teeth in it, pulsing with heart muscle. But for decades—for centuries even—teratomas have stood as perhaps the most intriguing evidence that somewhere in our cells lies the ability, dormant, and hidden, to grow an unlimited plethora of different tissues and cell types at any time. If only we could learn to harness it, the possibilities seemed limitless. After all, if we can grow a tumor composed of baby teeth, clumps of hair, and skin beneath our eyebrows or in our stomach, doesn't it stand to reason that our body, if properly instructed, could also grow us a new leg, even a new head, in the right place?

Yet for centuries, scientists, doctors, and philosophers have contemplated these kind of mysteries. How does a salamander regrow its tail? How can a teratoma appear in an adult man?

For that matter, how is it that Corporal Isaias Hernandez was able to regrow the muscle on the leg that doctors told him was forever destroyed and that 99.9999 percent of other doctors in the world would probably have told him he should just amputate?

For answers I've flown out to Pittsburgh to meet the man who discovered the pig bladder muscle regeneration technique in the mid-1980s, a trim, extroverted researcher named Stephen Badylak. I'm in a car next to Badylak, driving through the city on an icy winter day, and we're on the way back from a lecture he has just delivered to medical students and residents on the main campus of the University of Pittsburgh School of Medicine.

We pass gritty streets crammed with row houses, motor down a winding, tree-lined hill, and finally park in front of a gleaming office tower overlooking the city's Monongahela River and the steep wooded hills on the other side. Badylak's spacious labs are in a recently constructed, $21 million, five-story steel and glass facility, the headquarters of the McGowan Institute for Regenerative Medicine, one of the world's leading facilities in this burgeoning new field of research.

When we sit down in his office, I finally put to Badylak the question I have come to ask, making little effort to hide my skepticism. It's one he has answered a million times before, one that has alternately tormented him, fascinated him, guided him, and confounded him for years. I know all about myostatin, genetics, and even signaling agents like IGF-1. But how, I ask him, can you possibly regenerate muscle using a piece of *pig bladder*? Badylak smiles, wearily, with well-practiced patience and readily acknowledges that the idea sounds outlandish—so outlandish, in fact, that he spent years "reluctant to talk to clinicians about it."

"They didn't believe my results," he says. "Most people didn't believe it."

What has always made Badylak's findings so hard to believe to those who first encounter them is not just that he claims he could somehow regenerate human tissue with cells from another species—a procedure almost certain to trigger an immune response. He insists that his material can actually *transform* in a matter of months from what starts as a seemingly simple piece of a membrane used to collect urine, into whatever type of bodily tissue had been damaged—muscle, skin, or blood vessel. It's a claim that, until only recently, went far beyond the scope of accepted science.

No one was more skeptical than those who reviewed Badylak's initial application for funding to study this phenomenon when he first applied to the National Institutes of Health back in the 1980s and 1990s. Their response: "This is a crazy idea. This would never work. Who would ever try something like this?" How could this possibly work?

It didn't help that for years, Badylak didn't have a good answer for that question beyond the obvious fact that it seemed effective. No, he'd admit, he couldn't explain the mysterious mechanisms in play. He didn't understand them himself. He hadn't, in fact, even set out to regrow tissue in the first place.

As is so often the case in modern science, he discovered the treatment by accident. Unlike Sweeney, Badylak was more of a clinician than a back-lab scientist. His explorations, of course, would eventually lead him inexorably, just like the scientists we have met so far, on an ambitious quest to reverse-engineer the human body. But in the beginning, it was not paradigm-busting understanding he was after. In his initial approach, Badylak had more in common with a young Hugh Herr. He was a pragmatist, much more interested in solving an immediate problem with a jury-rigged replacement body part than in unraveling nature's secrets.

It all started, Badylak tells me with a mischievous twinkle in his eyes, with a "hair-brained" idea and a mutt named Rocky.

Back in 1987, Badylak was a newly minted hire at Purdue University working with a well-established vascular biomedical engineer named Leslie Geddes. The young Indiana native brought an unusual background to his new post. After college, he had pursued a degree in veterinary science at Purdue and practiced animal medicine until he realized that most pet owners couldn't afford the tests necessary to diagnose the conditions that fascinated him. Frustrated and worried he would grow bored, Badylak went back to Purdue to earn a Ph.D. in animal pathology. Then, after weighing teaching offers, he decided instead to go to medical school.

Badylak used his connections to help pay his way—he set up a lab in his home to diagnose cases of ferret lymphoma and dog breast cancer for former veterinary classmates who mailed him samples.

When Badylak finally began his research career, it was only natural that he would test a hypothesis geared to human patients on the animal subjects he knew best. In Pittsburgh, surgeons had pioneered an experimental technique on human heart patients, called "cardiomyoplasty," that removed a flap of a patient's back muscle, wrapped it around his or her failing heart, and then repeatedly stimulated it to help squeeze blood through the body.

But the procedure had a drawback—it relied on synthetic tubing to replace the aortic artery, which often triggered an immune response leading to inflammation and blood clots. It seemed to Badylak that if he could find a blood vessel substitute within a patient's own body, he could avoid an immune response and solve the problem.

One afternoon, Badylak sedated an affable mutt named Rocky, removed part of the dog's aorta, and replaced it with the one part of the body that most resembled the tubular structure of Rocky's blood vessels—a piece of his small intestine. Badylak didn't really expect Rocky to survive the night. But he figured that if Rocky hadn't bled out by morning, it would prove the intestine was sturdy enough to pass blood and worthy of further experiment.

It was, Badylak would later admit, the kind of "hare-brained," "out-of-the-box" experiment that you would probably never get past a university animal care and use committee today. Even back then, his third-year cardiovascular surgery resident called the idea "cruel" and "ridiculous" and refused to participate.

But when Badylak arrived for work the morning after Rocky's operation, he found the mutt wagging his tail, ready for breakfast. Badylak kept waiting for the dog to die, but days turned to weeks and Rocky continued to thrive.

"I didn't want to go in surgically and look because I wanted to see how long the intestine would hold," Badylak says.

Instead, Badylak repeated the procedure on fifteen other dogs. Six months later, Badylak finally cut one open. That, he recalls, is when "things got really weird." Badylak couldn't find the transplanted intestine. After checking and double-checking to make sure he had the right animal, Badylak placed a piece of tissue culled from the transplant target area under the microscope. What he saw astounded him.

"I was looking at something that wasn't supposed to happen," Badylak says. "It went against everything I had been taught in medical school."

Under the glass, he could still see traces of the sutures. But the intestinal tissue was gone. In its place, the aorta had grown back.

"Nobody would confuse an intestine and an aorta," Badylak says. "The microscopic picture is entirely different. I tried to get everybody I could think of to look at it. I kept asking, 'Am I seeing what I think I'm seeing?'"

Intestine is composed of soft, smooth, thinly lined walls, with hairlike projections known as villi. Aorta is thick, with the meaty, striated layers of the tissue that characterize muscle, those filaments and threads Lee Sweeney has spent so much time studying. Badylak examined several other dogs in the weeks that followed. And by varying the time allowed for the healing to take place, he was able to watch the intestinal tissue transform into vascular tissue again and again—proving it was no fluke.

Badylak had solved the mysterious "how" of Rocky's miraculous recovery—in fact, the dog would go on to live another eight years. But he now faced a much larger enigma, as he contemplated the "why."

Badylak came up with an initial theory. Maybe the body had a process of regenerating tissue that had been there all along, but the body's natural inflammatory immune response simply overrode it. That would explain why it didn't occur in operations using synthetic materials to replace the aorta—which had been plagued by inflammations.

To test his hypothesis, Badylak altered the experimental conditions in a way that, if he was correct, seemed guaranteed to generate an inflammatory response—he replaced the host dog's aorta with intestine from another dog. Badylak expected that the host dog's immune system would reject the material as foreign, leading to the inflammatory convergence of immune cells and an inhibition of the regenerative response.

But after Badylak sewed a piece of foreign intestine into a host dog's aorta, there was no inflammatory response whatsoever. Badylak repeated the experiment again—this time using cat intestine in a dog (oh the canine indignity!). He was sure that would produce the expected response. But he was once again astounded. The dog's body accepted it.

By now, Badylak knew he would be working with small intestines for a long time and he was going to need lots of them. So for his next experiment, he used intestine obtained from one of hundreds of pig slaughterhouses dotting the Indiana countryside surrounding Purdue. There would be no shortage of this material if it too worked.

Sure enough, the test dog was waiting for breakfast the day after the pig intestine operation, and many days later. Porcine entrails have been a staple in the doctor's labs ever since.

Badylak was now convinced that there was something about the small intestine that was also suppressing inflammation, in addition to promoting regeneration. He thought back to a bizarre paper on liver regeneration he'd heard about in a veterinary school pathology lecture. If you ate poison and destroyed all of the liver's cells, he seemed to recall, the organ could still regenerate itself—if the structural scaffolding of the organ, known as the "extracellular matrix"

(ECM), remained intact. Destroy this scaffolding, though, and the body responded with massive scar tissue. No regeneration occurred.

Badylak began stripping away layers of the intestine. And Badylak's suspicions were soon confirmed—when he stripped out all the living cells contained in the mucosa and outside muscle layers, leaving only a paper-thin ECM layer of small intestine connective tissues known as "submucosa," the regeneration worked even better.

Badylak was already dreaming about the medical applications for this mysterious material. And during that intoxicating initial period of discovery, running from 1987 to 1990, he pushed the limits of possible uses. He moved from the large aortic artery to large veins. Then he found the material worked on small veins. Finally, Badylak tried it on another part of the body altogether—he removed a chunk of a dog's Achilles' tendon and covered it with submucosa from a pig's intestine.

The normal response of any mammal to significant damage is to create scar tissue, instead of the more time-consuming process of regenerating what has been lost. Doing so likely had a clear evolutionary advantage: The body was quickly sealed off from lethal infections. Bacteria was kept out. We survived. But Badylak's dogs developed no scars on their Achilles' tendons, and thus no permanent limps. Instead they grew their entire tendons back.

To truly understand the cause of ECM's mysterious regenerative powers, Badylak would need to dig down to the cellular level and watch individual components of the tissue interacting in real time. But when he began to apply the full gamut of biochemistry to the structure of ECM, he found there wasn't much to go on.

The extracellular matrix was known to be the glue that holds tissue together, a cellular-level skeleton of sorts upon which the real machines of our biological processes—nerve, bone, and muscle—can plant themselves and get to work. It was composed of some of the body's most enormous proteins, structural building blocks known as collagen, laminin, and fibronectin, all woven together in an intricate, seemingly impregnable web to form a scaffold. Few had ever suggested the ECM was more than that.

But also woven into that matrix is another naturally occurring class of proteins known as growth factors, which can stimulate cellular growth and seemed promising. Using electron microscopy and biochemistry, Badylak and his team spent several years analyzing and publishing papers characterizing some of these proteins. They also identified the liver, the urinary bladder, the heart, and the esophagus as other potential sources of ECM-based materials. And they conducted experiments that revealed yet another important characteristic of ECM. When Badylak placed samples of known bacterium on a culture plate and added different pieces of the matrix, the bacterium failed to grow. Not only could Badylak's mysterious material suppress the body's natural immune response, it seemed; it also had its own antimicrobial properties that made such a response unnecessary.

Despite all these advances, however, Badylak readily admits that the true mechanism of ECM's healing power still eluded him when he sat down for a series of meetings in 1996 with representatives from a private company who had purchased the right to market the material from the university, and also the FDA to discuss initial testing of biologic scaffolds in humans. Despite the ambiguities, Badylak and his sponsors were able to win FDA approval through a backdoor tactic—the product was similar to other materials already approved for use as scaffolds or wound "patches," simple fillers not expected to offer any regenerative advantages.

After the FDA approved the material for testing, surgeons across the nation began using it on human patients for the first time. It was then that Badylak had his second epiphany—again, through a fortuitous accident that he could never have engineered on his own.

In 1999, Badylak visited one of these surgeons in Los Angeles. The surgeon, John Itamura, had implanted a scaffold into the shoulder of a patient, who returned eight weeks later in need of surgery for an unrelated problem. The lucky coincidence allowed the doctors to remove a sample from the area of the shoulder operation and check on the action early. A biopsy showed that the scaffolding had disappeared, as expected. But there was a surprise—a very active process was still transpiring at the surgery site, with an abnormal number of disparate cells flooding the area.

At first Badylak was puzzled. He knew that the scaffolding couldn't be causing the activity, because it had long since broken down. The cause, he realized, had to be the products left behind after the scaffolding had been broken down—molecules, perhaps, that had been lurking within the scaffolding all along, waiting to be released. Badylak began to comb the scientific literature for answers.

He quickly discovered that components called "cryptic peptides" could potentially explain much of ECM's unique phenomena. Researchers in other fields had previously identified these peptide fragments as components of larger parent molecules, and demonstrated that these peptide fragments could be released and activated when the larger parent molecules were degraded. These cryptic peptides were known to have potent antimicrobial effects and important signaling abilities, in some ways not unlike the muscle-building hormones we learned about in the last chapter.

"In the past almost everybody considered the extracellular matrix just the structural support that allowed you to stand up, and support weight, and hold things together," Badylak says. "But now we know it's almost just the opposite. It's primarily a collection of signaling proteins and information that is held within the structural molecules like collagen."

Badylak went back to the microscope and watched armies of tiny cells converge on the site of the broken-down ECM, apparently summoned by combinations of these signaling peptides. In their number and characteristics these new arrivals didn't look like muscle, nerve, or blood cells, but something altogether different and strange. They were unusually smooth, round cells. Badylak knew he was getting close.

And there was something about these cells that seemed familiar.

In April 1960, a young Canadian scientist named Ernest McCulloch piloted his beat-up Dodge through the streets of Toronto one Sunday morning to a lab at the Ontario Cancer Institute to check on his mice.

McCulloch, whose specialty was leukemia, had been closely following a series of electrifying experiments in the early 1950s that for the first time suggested the miraculous healing power of a new technique called a "bone marrow transplant."

By studying the devastating effects of nuclear weapons, scientists had learned that one of the primary results of radiation exposure appeared to be to disrupt the body's natural ability to replenish its supply of blood cells. This is a big problem because blood cells have among the most impressive turnover rate of any cell in the body, with each cell surviving just 120 days. The job of red blood cells is to carry oxygen and distribute it through our vasculature, a huge undertaking. We have *25 trillion* red blood cells alone in our bloodstream, which means we must replenish *roughly 2 to 3 million every second* to keep the oxygen supply flowing unabated.

Wound-clotting platelet blood cells, and infection-fighting white blood cells, meanwhile, usually last only a day.

Without interventions from scientists to help replenish their lost blood cells, mice in the lab exposed to radiation quickly lose the ability to deliver oxygen to the rest of the body, and to clot when wounded, and the animals die. But scientists discovered that if they replenished the marrow in the bones of the radiation-exposed mice with marrow from a healthy animal, miraculously the animal seemed to recover. Somehow the cells in bone marrow seemed to play an integral part in regenerating blood cells.

It didn't take long for enterprising researchers to glimpse the potential to treat cancer; if you exposed a mouse infused with tumors to enough radiation, it would kill both the tumors and the mouse's healthy cells in its bone marrow. Then you could simply replace the healthy cells via a bone marrow transplant.

Though the technique had already been demonstrated by others on that fateful Sunday, it was so new that many questions remained unanswered. What were the exact mechanisms at work—why was bone marrow so important? What was the rate at which cells exposed to different levels of radiation would die? How much transplanted bone marrow was needed to save an animal?

It was these questions and others that McCulloch and another young researcher, named James Till, had set out to answer when they irradiated scores of mice in their laboratory to kill the marrow in their bones, and replaced it with cells taken from the bone marrow of healthy mice. The two had come up with an elaborate experimental design aimed at precisely counting how many cells died, regrew, and survived.

Still, as McCulloch made his way to the lab to check on the mice that quiet Sunday in Toronto, there was little reason to believe that he was about to change science forever and lay the seeds for the new field of regenerative medicine. But a snap decision made at the lab would do precisely that.

After irradiating mice, McCulloch and Till had agreed to wait a number of weeks before removing the femurs and spleens and exhaustively examining the number and the health of the cells produced. By then any trace of the mysterious regeneration was long gone when they cut open most of the mice—though the impact of some form of regeneration was clear from the improved health of the mice who had undergone the transplants. But on that particular Sunday, only ten days had passed since the transplants, and McCulloch decided to sacrifice one of the mice early.

When he cut open the rear flank, he was amazed at what he

discovered. Clearly visible on the spleen, an organ that plays a central role in blood formation in mice, were clumps of large nodules that he hadn't seen in the mice given more time to mature. By meticulously counting the nodules, McCulloch was able to chart an unmistakable correlation between the number of marrow cells injected into the mice and the number of bumps on their spleens. Using radioactive markers, the team was soon able to demonstrate that each one of these lumps bulged with the precursors of red blood cells, white blood cells, and platelets. All of these precursors, it turned out, derived from a single cell that had been hidden amid the thousands of others infused into the bone marrow of the recipient mice.

These single cells were called "stem cells."*

McCulloch had glimpsed an elusive phantom that researchers had long before speculated existed, but had never before been able to isolate. He and Till had proved the existence of the stem cell, which they then defined. Stem cells, McCulloch wrote, are single undifferentiated cells that can multiply and give rise to differentiated cells. In the implanted mice, these healthy stem cells allowed the animals to regenerate the blood they needed to survive.

Stem cells, scientists would discover, explain the amazing abilities of cancerous teratomas—those "swollen monsters"—to differentiate into teeth and hair and skin. Stem cells are what allow a salamander to regenerate its limbs.

And in the shoulder tissue that surgeon John Itamura removed from his patient, these stem cells were regenerating the muscle—and they allowed Stephen Badylak to finally begin to understand the mysterious healing powers of the material he had stumbled on so many years before.

Peering into the microscope in the offices of that Los Ange-

* For more on the early history of stem cells, see Ann B. Parson, *The Proteus Effect: Stem Cells and Their Promise for Medicine* (Washington, DC: Joseph Henry Press, 2004).

les surgeon and watching those armies of unusually smooth, round cells converge on the muscle, Badylak realized he had found a way to summon armies of McCulloch and Till's stem cells to areas of devastated muscle—and somehow change the body's default healing mechanism. The particular type of cells he summoned are now known to reside in the bone marrow. Though they are not the most flexible of stem cells—that distinction belongs to less mature stem cells, such as those derived from an embryo, which can transform into any kind of tissue—the cells *are* recognized as among the utility players of the body, the armies of worker cells that can patch us up and produce many of the tissues we need when called upon. (One of the ways that Se-Jin Lee's myostatin works to inhibit muscle growth is by inhibiting stem cell activity.)

In 2003, Badylak took to the laboratory to prove his suspicions definitively. Taking a page from McCulloch and Till, Badylak first X-rayed mice to kill off all the stem cells in their bone marrow. Then he repopulated the bone with stem cells tagged with a fluorescent marker. When he removed a piece of mouse Achilles' tendon and added ECM, the fluorescent stem cells flooded into the area within days. Months later, some of these tagged cells were still present—implying that some of them had matured into regenerated tissue.

Ever since, researchers in Badylak's laboratories have been attempting to isolate the individual components in ECM that are capable of attracting the stem cells. They are doing so by using enzymes and detergents to break down ECM's parent molecules and separate the products into different "fractions" based on a number of properties, such as molecular weight. Researcher Janet Reing then runs an assay on the resulting fractions, using a device consisting of a row of discrete wells, all covered with filters that wall them off from a common channel. At the bottom of each channel she places a different ECM fraction. When she inserts different stem cell types into the common channel, she can now observe which of the discrete fractions attract the stem cells with greatest affinity.

Throughout the 2000s, Reing and the other researchers have gradually winnowed down the size of these fractions, from soups containing thousands of molecules into ever smaller and more specific segments, homing in on the individual peptides. Badylak and his team now believe some of these peptides are also responsible for suppressing the natural scarring response, which helped us survive before the dawn of modern medicine, when a simple injury could lead to death by infection.

"Scarring makes sense from an evolutionary perspective," says Ricardo Londono, a M.D.-Ph.D. candidate in Badylak's lab. "Before modern medicine, every single time we got injured for the most part it meant death because of loss of blood and infection. Throughout the millions of years of evolution, sealing up a wound became a top priority. When you make a scar it's just a quick-and-dirty way to patch it up."

Londono has spent the last five years focused on understanding the early immune and stem cell response to implanted ECM-based biomaterials. Once tissue gets injured, he notes, it's too late to start generating from scratch the cellular and molecular signals that would be required for tissue regeneration and repair to occur. Protein expression and signal production can take hours, even days, he notes. Nature's elegant solution, therefore, is to create the signals beforehand, but to encrypt them.

"It's almost like when you give the nuclear codes to the submarine," Londono says. "The orders are there, but they're encrypted and they're not allowed to turn the key until it's actually needed. These signals are hidden in the ECM in the form of cryptic peptides, and the way in which they are encrypted is by making them part of the larger molecules and making their active site physically unavailable to nearby cells to read."

Blow off a big chunk of the leg, as that artillery shell did to Corporal Isaias Hernandez that day back in western Iraq, and the encrypted signals are no longer present to be decrypted. What Badylak discovered is that you could add these signals back into the injury site in the form of an ECM-based biomaterial and the body would degrade the ECM, decode the encrypted message contained within it, and then summon the stem cells to do their work.

Stem cells and other cells like them explain a lot and not just the mysterious healing Badylak witnessed in Rocky. Studies have shown that one of the reasons myostatin—targeted by Sweeney, as we saw in the last chapter—can inhibit muscle growth is that it can keep stem cells in a "quiescent" state and inhibit their self-renewal. IGF-1, on the other hand, which Sweeney used so effectively to make his Schwarzenegger mice and dogs, helps promote stem cell activities.

What's more, stem cells are capable of building far more than just muscle. Trace back any kind of body tissue far enough and you'll find stem cells. Stem cells form our brains, our hearts, our blood, and our teeth. They explain how we grow bones.

But how is it that a stem cell knows what to do? What determines whether it will become a piece of Rocky's intestine, or a piece of new muscle? How does it turn into part of an internal organ? Or a new person for that matter? And how far can we take this discovery?

That's a question many in the burgeoning field of bioengineering are actively pursuing. One such researcher is Gordana Vunjak-Novakovic, a Serbian-born expert who is also pushing the very edges of what's known about how the body builds, heals, and regenerates.

Back in the 1980s, around the time that Badylak conducted his first experiment on Rocky, Vunjak-Novakovic arrived at MIT on a Fulbright Scholarship from her native Serbia to do work in the laboratory of another pioneer in regenerative medicine, a man whose name would come to be synonymous with the field later known as "tissue engineering": Robert Langer.

The experiments that Vunjak-Novakovic would pursue along with Langer and others in his lab would reveal a series of important insights about the body's own natural healing responses, and some of the other internal signals that direct them. By understanding these signals, Vunjak-Novakovic and her collaborators (Badylak among them) are helping science move closer to the long-elusive goal of having control over tissue regeneration.

They are discovering not just how to summon stem cells to the site of a wound—as Badylak has done—but also how to isolate them and experiment with them outside the body. They are learning how to direct the type of tissue they become, how to take control of the elusive phantoms that produce those cancerous masses of hair, teeth, and skin—the swollen monsters. And they are producing remarkable products—not just muscle, but skin, cartilage, and bones.

When I visit Vunjak-Novakovic in her twelfth-floor office at Columbia University Medical Center's Vanderbilt Clinic on Manhattan's 168th Street, she leads me into a refrigerated room full of vials. Then she reaches into a cabinet and pulls out a pieces of heart made in a lab. It's uncanny. The tissue, on its own, appears to be beating.

"Stem cells take their cues from the nutrients they receive, the intensity of electrical impulses they feel, the oxygen they get, and the movement they experience," Vunjak-Novakovic says. "All these factors, in addition to the physical dimensions of their surroundings, indicate to the stem cells what part of the body they're in. We need to create an artificial environment that mimics all of that enough to 'instruct' the cells what exactly to do."

While Badylak's background as a veterinarian placed him perfectly as an experimental surgeon in the burgeoning field of tissue engineers, Vunjak-Novakovic's academic specialty prepared her to take a leading role in another part of the frontier: building these artificial environments and finding ways to control them.

Originally, when Vunjak-Novakovic was working toward a

Ph.D. in chemical engineering in the early 1980s at the University of Belgrade, the prospect of growing body parts never occurred to her. She was interested instead in attempting to understand the forces and motions created by the intermingling of gas bubbles and tiny solid particles in liquids. Her research involved mathematical modeling and experiments in enclosed reactors, and was applicable, most obviously, to industries that rely on fermentation, like food production and the manufacture of penicillin and other antibiotics. It was a pursuit that required her to build reactors where the chemical interactions could be meticulously emulated and controlled.

As a young faculty member at the University of Belgrade, Vunjak-Novakovic soon became captivated by the chemical interactions that take place among molecules within living organisms, an interest that came at a fortuitous time. In 1986, during her Fulbright studies at MIT, she caught the attention of Langer. He was trying to detoxify human blood in patients and was looking for someone to create new machines to selectively remove drugs from blood.

After Vunjak-Novakovic returned to Belgrade, she made biannual trips back to Boston, in addition to remaining in regular contact with Langer and his other collaborators. During one such visit, in 1991, ethnic tensions in her homeland boiled over into war. "It became clear that it would be good to leave Yugoslavia," says Vunjak-Novakovic. Eventually the situation deteriorated so much back home that when concerned colleagues at MIT learned in 1993 that her visa was about to expire, they lobbied successfully to get her a permanent position that allowed her to stay in the United States with her husband and young son.

Around the same time, Langer announced he had received a grant to do something called "tissue engineering" and asked if Vunjak-Novakovic would like to join the project.

Langer was about to develop some of the field's most important lab techniques. While Badylak's contributions relied on signaling agents, one of Langer's main contributions was to demonstrate that the shape, architecture, and the degradation of materials inserted into a wound site could also play a key role in regeneration. He created three-dimensional molds, or "scaffolds," that could be seeded with regenerative cells and then put safely into a person's body. These scaffolds guided the development of the emerging tissue while the synthetic materials biodegraded at the same time.

When Vunjak-Novakovic started to work at MIT in 1993, her first project was to try to create cartilage, the flexible connective tissue that makes up the nose and the ears and is found between many joints. Compared to bone, which is harder and less flexible, and muscle, which is softer and stretches more, cartilage seemed an easier reach for regeneration. The gel-like tissue consists of a single cell type, is far less structurally complex, and is devoid of the blood vessels that are critical for survival of bones and muscles. Collaborating with another young scientist, Lisa Freed, she worked to find the way to grow this "simple" tissue.

At the time, tissue engineers experimenting with the cultivation of stem cells outside the body believed their primary tool lay in feeding it the right mix of nutrients, minerals, and proteins as the cells grew and matured. Even slight variations in the nutrient soup they injected into their molds, they learned, had a profound effect. A little extra calcium, for instance, would signal to the stem cells to develop into bone.

Vunjak-Novakovic, however, suspected there were other factors at work. She was reading a lot about mechanobiology at the time and was fascinated to discover that many physiological systems—genetic, molecular, electrical, and mechanical—are interconnected in surprising ways. In particular, she noted that people immobilized in hospital beds for long periods of time often experience a weakening of their bones and cartilage. It seemed that physical movement was somehow essential to the upkeep of these tissues. Was it possible, she wondered, that developing cells are also sensitive to movement? How exactly could a mechanical phenomenon—something involving forces or displacement (or the lack of it)—affect the tissues of bone on a cellular level? Vunjak-Novakovic, her collaborator Freed, and their students began slowly rotating the vessels containing their cellular colonies on biomaterial scaffolds to test the hypothesis, and quickly obtained exciting results. The movement did indeed seem to aid in the growth of the cells—in unexpected ways.

"We discovered that if you have physical factor, it helps, and if you have growth factor it helps," she says. "But if you use them together in a clever way, you have this synergistic effect. Two and two is not four; it's nine. You get a tremendous improvement if you use them interactively."

"The improvement in its structural integrity," she says, "was beyond what we'd imagined."

Even so, it wasn't until several years later that Vunjak-Novakovic and her colleagues would fully understand the dynamics at work. They would discover the phenomenon in an unexpected environment: outer space.

In 1996, scientists at NASA decided to conduct the first experiments with tissue engineering in space on the International Space Station. Scientists at MIT, which had a long history of collaboration with the space program, were obvious candidates for recruitment. And the pioneering work conducted by Vunjak-Novakovic and Freed, confined as it was to carefully controlled and easily contained bioreactors, seemed ideal.

Since NASA did not know when it would launch, or when they would be able to bring the samples back, Vunjak-Novakovic and Freed designed the sturdiest experiment they could think of. Once their proposal was accepted, they loaded up pieces of bioengineered cartilage into a bioreactor, along with the nutrient solution that would be mixed with oxygen and perfused over the cell culture every day, and fit it all into a box about the size of a small microwave oven. Then they sent it up to space.

When the box finally returned after four and a half months in space, Vunjak-Novakovic and her colleagues fully expected to find a cell culture that demonstrated superior growth, thanks to the lack of gravity, which would have added resistance to the gentle rotation of the bioreactor. After all, it seemed that these conditions would mimic the conditions of the embryo, allowing the cells to float freely in suspension.

But Vunjak-Novakovic and Freed were shocked to find exactly the opposite. The cells hadn't done well at all—they had done far worse. It was then that they realized it was not a lack of movement that prompted the tissues of convalescing hospital patients to atrophy. It was a lack of sufficient force—the lack of sufficient mechanical load, generated by the combination of muscle movement and gravity, pushing downward on the cells.

"At the time, the mantra was that 'everything is better in space—there is no gravity and things function better,'" Vunjak-Novakovic says. "And we found exactly the opposite. It was an interesting result because it also showed why astronauts had a number of physiological problems, including a lot of bone loss and cartilage loss."

The findings led to a high-profile scientific paper, and also to a major innovation in tissue-engineering technology that vastly improved the quality of bone and cartilage Vunjak-Novakovic is able to grow. She developed a plunger that would gently press down on the tissue and the soup of chemicals washing over it. Further experimentation revealed that intermittent compression worked best.

"We don't run and walk all day," Vunjak-Novakovic explains. "We walk and sit and then again walk and sit."

Like so many bioengineers before her, including Hugh Herr, who would later work just across campus and would eventually open a bioengineering institute with her erstwhile colleague Robert Langer, Vunjak-Novakovic was coming to realize that the most effective approach was one that was "biomimetic," replicating the conditions in nature.

"The whole field was driven by molecular factors for a very long time," she says. "And then the biomaterials came into the picture and it was thought that an ideal biomaterial should be inert, doing nothing. It took a lot of time and effort of very many people to come to the point that we understand that biomaterials need to tell the cells what to do. Because the cells touch it, pull on it, press it, and feel it. We started to think that an ideal biomaterial is a biomaterial that looks and works like a native tissue matrix."

It was a lesson that Badylak would eventually learn with his dogs. If he restricted their movements too much after implanting ECM, it was impossible to grow back an Achilles' tendon. The cellular matrix, even when placed in their bodies, needed exposure to naturalistic conditions to work its magic.

The lesson was also taken to heart by Laura Niklason, a postdoc at MIT back in the 1990s who shared an office with Vunjak-Novakovic in Langer's lab. Niklason is a pioneer in growing arteries. Initially her primary tool was to try to coax stem cells to grow into arteries by replicating the chemical soup that surrounded the embryo during various phases of development. But when she experimented with different types of scaffolds that held the stem cells in place, she too made a surprising discovery.

"The belief at the time was that, if you're going to grow an artery, you need a very strong polymer that's not going to degrade," she says. "Because arteries have to withstand physical forces, and if they rupture, it's bad for the patient, right?"

Yet when Niklason took this approach, the arteries that resulted were weak and looked nothing like the real thing. When she began experimenting with scaffolds of different properties that degraded at different rates, she was surprised to find that the scaffolds that degraded the fastest actually created the strongest, most realistic arteries. Once again the solution was to look toward nature and emulate it. Without exposure to the physical forces unleashed when scaffolds degraded quickly, the arteries missed out on crucial environmental cues apparently needed by their cells to appropriately adjust and calibrate arterial strength.

Controlling the environment of stem cells is far more essential than scientists originally realized, Vunjak-Novakovic says.

"All these factors, in addition to the physical dimensions of their surroundings, indicate to the stem cells what part of the body they're in," she says. "We need to create an artificial environment that mimics all of that, and directs the cells to form the right types of tissues in the right place and right time."

In 2005, Vunjak-Novakovic arrived at Columbia University and took aim at the next frontier—the challenge of heart tissue. To Vunjak-Novakovic, the idea of engineering healthy heart tissue seemed among the most challenging pursuits. Cardiac cells start to die within fifteen to twenty minutes if they are deprived of oxygen, which is what occurs when arterial blockages lead to heart attacks. Unlike some other tissues in the body, heart tissue is incapable of healing on its own. Instead, the body will paste over the dead area with scar tissue, leaving a permanent obstruction devoid of activity and that reduces pumping capacity. This scar not only weakens the rest of the body, it leads the heart to thin and enlarge, and eventually causes it to fail.

If she could develop tissue capable of replacing the dead cells, Vunjak-Novakovic reasoned, it could pick up the slack and help prevent expansion. Further, by allowing new blood vessels to grow into the area, it could bring in regenerative cells, nutrients, and oxygen capable of taking over the dead tissue, clearing away blockades and making way for new growth. The possibility of this miraculous fix had dawned on Vunjak-Novakovic one morning in 2001. By now she knew where to start. Just as she did with cartilage, Vunjak-Novakovic would look to nature for refinements she might add to her bioreactor to coax heart cells into growing.

One of the fascinating characteristics about heart tissue to her is that it is the first organ in the body that starts to function. It actually starts beating just three weeks into embryotic development, when electrical signals from the top of the heart spread over cells all the way to the bottom, depolarizing cell membranes and causing the cells to twitch. In 2001, Vunjak-Novakovic wondered what would happen if she applied a pulse from a clinical pacemaker to cardiac muscle cells. Might this prove to be the factor that would stimulate their growth? To find out, she and then–graduate student Milica Radisic placed the heart cells of a rat onto a soft, elastic scaffolding material, added a fluid medium, and began to stimulate it intermittently with the pacemaker.

One morning about a week later, Vunjak-Novakovic walked into her lab, removed the cells from the pacemaker, and placed them under a microscope. Just as she bent over to look, one of her postdocs walked in, slamming the heavy lab doors behind her.

"Wait," Vunjak-Novakovic said. "You disturbed everything. The whole system is shaking."

She rose, locked the door, headed back to the microscope, looked again, and realized the cells were still moving. They were not shaking at all—they were beating unassisted.

"Within a week, they started from something totally disorganized and diffused—one cell here, one cell there—and they made an organized community that started to function," Vunjak-Novakovic recalls.

In fact, the tissue looked so similar to native heart tissue that at first she was convinced she had somehow mixed the engineered sample up with the native heart tissue control group. So she repeated the experiment and sent it off to a colleague in Switzerland for independent analysis. He could not distinguish between the two cardiac samples.

"The bar is lower than you originally thought," Vunjak-Novakovic realized after the pacemaker experiment. "You realize you don't need to do so much" to get the cells to behave the way you want. All you need to do is provide them the cues they are accustomed to receiving in the body. Once you do, you "activate the genetic machinery of the cell, and the cell would recognize something in their environment.

"Depending on the combination of these factors, which very often change in space and time, certain genes will be turned up, other genes will be turned down," she says. "The outcome may range from super-favorable, the cell doing its job of tissue regeneration, to very unfavorable, the cells dying."

When you think about it then, the approach of Badylak when he regenerates muscle is not that different from that of Vunjak-Novakovic and Laura Niklason. To direct the activity of stem cells, Niklason and Vunjak-Novakovic rely on man-made "bioreactors" that allow them to carefully control the intermittent application of force and the characteristics of the chemical soup with its myriad signaling agents and nutrients. Badylak simply inserts his magical scaffolds into the wound site and lets the body do the work for him.

It's easy to see, through the lens of what we now know, that the very same factors at work in man-made bioreactors are at work in Badylak's natural-made bioreactor of choice. Corporal Isaias Hernandez's thigh muscle grew back not just because Badylak inserted biologic scaffolds into the remaining tissue, and not just because that scaffolding then broke down, released signaling agents, and summoned stem cells to the site. The muscle grew back because Hernandez was grunting and sweating through physical therapy every day. Every time Hernandez allowed his weight to come down on those stem cells, it sent a signal—the same signal that Vunjak-Novakovic conveyed artificially in her bioreactor with the plunger that she created to gently press down on the tissue to coax it to become bone or cartilage.

One of the holy grails in the field of regenerative medicine is the ability to engineer whole organs in all their complexity, not just pieces of them.

Laura Niklason is one of those pushing this boundary. During a visit, I followed one of her postdocs into a refrigerated closet in her Yale University laboratories. He reached out to a shelf and took down a jar. Unlike the amorphous piece of heart muscle Vunjak-Novakovic had showed me, there was no mistaking what was floating inside this container. It was a perfectly preserved pair of rat lungs, taken from an actual animal and "decellularized."

Like those who are engineering simpler tissue, Niklason relies on physical forces and a chemical soup to replicate the native environment of the organ and coax stem cells to mature into the kind of tissue she desires when manufacturing lungs. But she came to believe early in her efforts that science did not yet offer the technology to construct an artificial scaffolding detailed enough to emulate the shape and architecture of a real lung, a complex structure as labyrinthine as a Minotaur's maze. We inhale air through the human trachea, a single passage, which quickly branches into smaller offshoots, which in turn project branches of their own. There are, in fact, twenty-three generations of branching in the airways of our lungs, and hundreds of millions of air sacs 200 microns in diameter, each one filled with capillaries that absorb oxygen into the blood.

"If you try to make a polymer that's got all that built in . . . ," Niklason tells me, trailing off and scrunching up her face as she contemplates the magnitude of the challenge. "The technology isn't there. Doesn't exist. Full stop. Can't do it."

Instead, Niklason relies on nature to do it for her. After taking lungs out of a dead donor, she soaks them in a combination of detergents and strong salt solutions to wash away all of the cellular components of the lung most likely to trigger an immune response when placed in a new body. What's left behind is a raw scaffold, the same kind of fibrous material Badylak uses when regenerating muscle, a structure whose biochemical components are largely the same across individuals and species. Unlike Badylak, however, what's important for Niklason in the early steps is the scaffold's complex architecture, its exact shape. Once the scaffold has been cleansed, she perfuses it with stem cells and places it in a bioreactor that aims to replicate the conditions a normal lung is exposed to inside the body.

"Blood perfuses through our lungs," she explains. "So, we had a setup so that we could perfuse the lung tissues and allow them to breathe as well, because breathing is important for lung development. Then we spent a lot of time working on the soup, too. So, it's scaffold, and bioreactor, and soup."

Niklason is not yet ready to test her lungs out on human patients. She notes that so far, no one has implanted such artificial lungs even in a rat for more than a day or two. The engineering requirements for a human, she notes, must be flawless, since the human recipient will likely live for many years. She calls up the cautionary tale of genetic therapy, which killed young Jesse Gelsinger and almost wrecked the career of Sweeney's former collaborator Jim Wilson.

"It's like building the Brooklyn Bridge," she says. "You have to decide how long it's going to be, how wide it's going to be, how much weight it can sustain. Can it withstand wind shear, changes in temperature? It's got to fulfill those criteria before you let people drive their cars across. Otherwise they crash into the East River."

Niklason's former office mate, Vunjak-Novakovic, is also working on ways to regenerate lungs. But she is taking a different approach. There are ten times the number of patients who need lung transplants as there are lungs from organ donors, she notes. What's more, about 40 percent of the donated lungs are rejected due to defects or damage during transport.

Instead of building an entirely new lung, Vunjak-Novakovic and her lung team place these damaged lungs in a humidified machine she calls "Deep Breath," which perfuses oxygenated blood, or blood substitutes infused with nutrients and oxygen, through the organ to simulate the real-world conditions of breathing. Then she and her team look for damaged areas and build healthy centers of tissue throughout the organ by adding stem cells, derived from the patient's own body, and using them to regenerate discrete pockets of lung tissue. They think that just seeding a few new colonies of stem cells could improve lung function to the point where an organ that would otherwise be rejected is considered functional enough for a transplant—and presumably to save someone's life.

"We look for the worst possible areas and try to fix them instead of removing everything and repopulating everything," Vunjak-Novakovic says. "The surgeons tell us that in most cases, if we can improve lung function by ten, fifteen, twenty percent we may be able to reach the bar for implantation and the body will do the rest, rather than starting from zero."

By 2007, Badylak's growing list of publications had created a stir in the fast-growing field of regenerative medicine, and his professional reputation was flourishing. To the outside world, however, the researcher remained largely unknown until, that year, an odd confluence of events catapulted him into the public eye.

A number of years before, Badylak had met a Boston-based surgeon named Alan Spievack, who approached Badylak after the latter delivered a lecture on ECM at an orthopedic conference in Atlanta. As an undergrad at Kenyon College in Ohio in the 1950s, Spievack had performed amputations on salamanders and studied the way the creatures regenerated their limbs. He went on to a long, successful career as a surgeon. But Badylak's talk rekindled Spievack's fascination with tissue regeneration, and Spievack persuaded Badylak to accompany him for a cup of coffee. Soon after, Spievack visited Badylak's lab and joined the growing number of scientists who had begun pursuing their own research on ECM.

By 2007, Spievack had coauthored several papers with Badylak, and even gone on to found a company called ACell to market his own special formula of the powder. So one afternoon that year when Spievack, by then seventy-three, got a call from his younger brother, Lee, seeking his medical advice, he knew exactly what to do.

Lee, a tough-as-nails Vietnam vet, and a modeler all his life, had gotten a job in a hobby shop after retiring. That day, he'd helped a customer fix a large model airplane, and had been letting the plane idle nearby as they conferred, when he pointed with his index finger a little too close to the whirring propeller blade. The propeller had neatly severed Lee's finger about a half inch from the tip. By the time he called Alan, Lee had stanched the bleeding, searched in vain for the severed tip, been to the hospital, and even made an appointment to see a hand surgeon the following Monday.

Lee's hand surgeon wanted to take a piece of skin from his thigh and sew it on the end of his finger. But Alan had a better idea. He told his brother to cancel the appointment.

"They got extremely upset," Lee recalls. "The receptionist said, 'You're going to get infections! You're going to have all kinds of problems. You are probably going to lose your hand!'"

Alan told Lee he was sending him a vial of powder derived from ECM and instructed him on how to apply it to his finger. Miraculously, within a couple of weeks the finger had grown back.

"The fingernail seems to grow faster than the rest of my body, because this fingernail is about four and a half years old and the rest is seventy-two years old," Lee told me a while back, explaining that he noticed he has to clip it more often than other fingernails. "The finger is hard on the end, but I have complete feeling and complete mobility on it."

When news got out that Lee had regenerated his fingertip with a mysterious powder he called "pixie dust"—and had the graphic pictures displaying the regenerative process to prove it—a media frenzy erupted. The stories and the photos sparked the imagination of amputation victims around the world, including Corporal Isaias Hernandez. Years later, Badylak still gets several e-mails a day asking about "pixie dust." (Alan Spievack did not get to share in much of the glory; he died of cancer in May 2008.) Badylak's regenerative work, it seemed, had finally gone thoroughly mainstream.

The story of Lee Spievack raises a tantalizing question. So far, researchers have demonstrated they can regrow muscle, skin, tendons, even organs. But what about complex body parts consisting of multiple tissues? Might it be possible someday to, for instance, regrow Hugh Herr's amputated leg?

It's a question both Badylak and Vunjak-Novakovic have pursued independently. They have done so on different occasions by teaming up with another emerging figure in the field named David Kaplan, chair of the Tufts University Department of Biomedical Engineering. Kaplan has been working with a watertight sleeve made from materials such as silicon, rubber, and silk. The device, which he calls a "biodome," can be placed directly over an amputation site, allowing his team to carefully control the conditions under which wound healing—and eventually, they hope, full-scale regeneration—takes place.

Working with mice and adult frogs that normally wouldn't regenerate amputated digits, Kaplan and his collaborators are using the sleeve to create the kind of protected, nutrient-rich, aquatic, life-sustaining environment you find surrounding an embryo. Among other things, Kaplan has added some of the peptides isolated in Badylak's lab to suppress inflammation and scarring, perfected ways of carefully controlling humidity to prevent the wound from drying up and dying, developed new kinds of gels to control external pressure, and experimented with a number of other external cues in the hopes of inducing full finger regrowth.

But perhaps the most powerful and potentially transformative factor Kaplan is adding to his biodome to induce limb regrowth isn't coming from the labs of Badylak or Vunjak-Novakovic but rather from Michael Levin, an intense biologist, who directs the Tufts Center for Regenerative and Developmental Biology. Levin believes the most important signals needed to grow a new hand, a leg, or even a head are encoded in electrical voltages present across each cell membrane.

Using genetic engineering and pharmacology, Levin is controlling the activity of special proteins (called "ion channels") in the cells' outer surfaces. When present these extra proteins, which are hollow in the center, allow more ions with positive or negative charges to flood into or out of the interior of the cell. This changes the voltage potential across the boundary of the cell. When we alter the natural differences among cells' electric potentials through the body, Levin argues, it can have powerful consequences, by turning on and off key genes not just in the cells with the extra receptors but in large numbers of cells nearby that are exquisitely sensitive to the electrical signals they receive from all the cells around them.

It's more than just a theory. Using these methods, Levin's lab has induced a frog to grow six legs and worms to grow two heads, and turned part of the gut of a tadpole into an eye. He has spurred the complete regeneration of severed tails that normally wouldn't grow back, complete with spinal cords, and reprogrammed tumors back into normal tissue. Levin believes his technique may one day provide the key to actually regrowing a human limb.

Electricity, he argues, is one of the universal signals the body uses to control how large-scale assemblies of cells work together to form organs and body parts of different shapes and sizes in different places. By decoding and tweaking these electrical signals, Levin hopes eventually to be able to take control of this process.

Using this technique along with the biodome to change electrical signals at the amputation site, Levin and Kaplan most recently demonstrated they could induce an adult frog to begin to regrow part of a severed limb, complete with bone and other tissue.

"The bioengineering approach to regrowing a limb would be

to micromanage the process—to say 'I can make a bunch of different cell types and now I'm going to arrange them all together in the pattern required to make a functional appendage,'" Levin says. "But for something like a limb that is never going to work. It's way too complex. What we're trying to do is understand in the native situation, the cells themselves, how do they decide what shape they should be making, how do they make it, and how do they know when to stop?"

Kaplan and Levin are now attempting limb regeneration in mice, which is a larger challenge than in frogs because mice are warm blooded. Warm-blooded animals have a much higher blood pressure, so the risk of bleeding out if the amputation wound is not immediately peppered over with a scab is substantial. Warm-blooded animals also have faster metabolisms, which makes the animal more likely to quickly succumb to infection. Thus the body's natural response is to launch a major inflammatory attack, which must also be suppressed. All of these factors make Kaplan's biodome, now in its fifth iteration, essential.

I must admit, I was skeptical during my initial visit to Levin's and Kaplan's labs. But finding the solution to limb regeneration in mammals, they both argue emphatically, is not a question of if, but a simple a question of "how much time is it going to take?"

"If somebody says they're going to fly to the moon on their own power, you can say, 'That's impossible,'" Levin told me. "'There are no examples of that.' Right? Okay, fine. But I fail to see how anybody can say that limb regeneration is impossible, because there are animals that do it. Period. It's clearly possible because we see salamanders doing it."

One intermediate approach might be to give the cells a head start, rather than attempting to induce full-scale regeneration at the site of the wound. In a separate study, Kaplan and Levin have teamed up with Vunjak-Novakovic to create a "logistic template" to guide the stem cells to grow into different types of tissues.

"It's like a Lego: You combine pieces, and then each piece is its own biological identity," Vunjak-Novakovic explains. "It has to have the shape, internal structure, composition, and mechanics. You can use a three-D printer or you can use depopulated scaffolds, without the original cells—that you machine into the necessary shapes."

Vunjak-Novakovic has already demonstrated she can regenerate pieces of a body with multiple tissue types. She points to the late movie critic Roger Ebert as someone who might have benefited had her technologies been developed a few years ago. (Ebert had a large portion of his jaw removed after cancer.) She has demonstrated that if you can take an image of the intact jaw before surgery—or by imaging the jaw on the other side—it's possible to make a 3-D mirror image on a computer. This 3-D image can then be used to make blocks of different materials needed to form a multi-tissue scaffold—machining the bone and cartilage separately, combining them, and then fitting them like Legos into the space where a missing jaw once was.

"This is right now becoming possible," Vunjak-Novakovic says. "We have done preclinical studies in a large animal mode. We got great data, so we started a company and are moving towards clinical trials."

Building an entirely new hand remains "one of the big goals."

"It's a very large goal to shoot for, but I think absolutely it's doable," Levin says. "It is going to work in the end."

On a balmy spring day in 2014, I fly down to Delray Beach, Florida, and head to the office of trauma surgeon Dr. Eugenio Rodriguez, whose medical exploits in recent years have garnered him headlines on the local news and in the newspaper. Rodriguez is not waiting for the results of Badylak's trials to apply the new technologies.

In 2011, one of Rodriguez's patient brought him an article in a mainstream magazine about Badylak's technique and asked the doctor to try it out on him. Since then, Rodriguez informed me on the phone when I first called him, he has used it literally "hundreds of times."

Certainly as I take in the scenery while driving from the airport to Rodriguez's office, it seems to me the doctor's practice is located in the ideal place to find patients for naturalistic experiments on human regeneration. It's a blindingly bright Florida day and I am close to the beach. When I pull up at a stoplight a couple of blocks away from my destination and look to my left, I notice that the car next to me is trailing the kind of wheeled metal platform you'd normally see carrying a jet-ski. Instead of a jet-ski, however, the driver has strapped a motorized four-wheel, geriatric utility scooter onto the back, and hooked a shiny metal walker onto its wire basket.

The cars I pass on the road, I realize, are all roomy American-made sedans, driven by stooped elderly men and women in their eighties.

After reading the article suggested by his patient's father, Rodriguez contacted Alan Spievack's old company, ACell (where Badylak now serves as chief science officer), and ordered some ECM.

When I arrive at Rodriguez's office he leads me down a long hallway and into an examination room. Sitting on a doctor's chair is Yancy Morales, a baby-faced twenty-one-year-old with a wispy mustache, wearing baggy red shorts and a Miami Heat jersey. After we shake hands, Morales extends his right leg straight out and motions to a nasty scar that runs for several inches on the inside of his leg above the knee. Then he points to a spot just above where the scar ends at the center of his thigh. It's the spot, he tells me, where a doctor drew in felt-tipped pen the line where he intended to amputate.

Morales had arrived at the hospital a couple of days earlier with his leg split open "like a butterfly and the bone was sticking out," after a car accident. After several surgeries, the doctor sent a nurse in to inform Morales that there was nothing they could do to save his leg.

"I'm picturing myself without my right leg," Yancy says. "It just kept on coming at me because my leg was going to be gone, you know? I'm not going to lie, I did cry."

One of the nurses had seen Rodriguez on the local news. So they tracked the doctor down. Rodriguez examined Morales. Yes, he told him, he believed he could save the leg.

After growing back some of the lost tissue in Yancy's leg with ECM, Rodriguez took skin from Yancy's upper thigh and grafted it into the wound to fill the remainder of the hole. Yancy has been assiduously rehabbing it ever since, putting loads on the new tissue to build it back up. He says it still hurts and swells up some days if he works it too hard, but there was no mistaking the gratitude he demonstrated toward Rodriguez.

"He saved my leg!" he says. "They were going to take my leg!"

After I spoke with Yancy, Rodriguez led me into another examination room down the hall. Sitting on a chair, surrounded by family members, is Mercedes Soto, a thirty-five-year-old housewife from Caracas, Venezuela, who has a second home in Miami. In 2013, Soto visited Florida when she was twenty-two weeks pregnant. She planned to have the baby in the United States. Two weeks later, Soto came down with an infection, miscarried, hemorrhaged, and went into septic shock.

The doctors in Miami put her in a medically induced coma and managed to keep her alive by using machines to keep the blood pumping to all her major organs. But the circulation to her extremities was severely limited. When she finally came to, one of her feet and parts of her fingers were black with gangrene. The vascular surgeon informed Soto that she intended to amputate the whole foot, at the ankle. But Soto's neighbor in Miami happened to work as a nurse in Delray Beach and told her about Rodriguez.

Sitting on a doctor's chair, Soto shows me a picture on her iPhone of the way her foot used to look. The top of the toes and the front of the foot are charcoal black. The bottom is bloated and covered in what looks like dark green, gangrenous bubbles. Now the foot is almost back to normal, though it was too late to save the toes. Soto's foot today ends in a rectangular wedge, with red and yellow skin held together with staples. Rodriguez has been attempting to grow back the toes using ECM, but so far, the results have been less than he might have hoped. ECM did, however, succeed in growing back a portion of her fingers, which Rodriguez had also amputated.

"The index finger grew back very significantly; the nail came back and all," he says. "The toes were not as impressive. There was more damage to there. But at least we did not have to amputate the foot."

To my untrained eye, the toes look like they have gone through a meat grinder; the digits have been almost completely cut off. But when Soto motions to the place where doctors told her they had intended to amputate, the miracle is clear. She is pointing to the bottom of her ankle.

Having spent so much time speaking with Hugh Herr, and re-creating the events of that ill-fated hike with Jeff Batzer into the snowy wilds of New Hampshire, I find it easy to imagine alternative endings to the stories Morales and Soto share with me. The vividness with which both Jeff Batzer and Hugh Herr had related what it was like in those early days to lose a limb—told years after the fact—had stayed with me. And a strong sense of déjà vu hits me each time one of the patients in that Florida clinic recalls that chilling moment when a solemn-faced doctor or nurse strode up to their hospital bed and informed them of the need to take a limb. Time stops when news like that is delivered.

Yet how different the stories of Morales and other patients I met that day turned out to be. And how different the lives of future accident victims might be if the technology of regenerative medicine continues apace. It is still, of course, the early days of regenerative medicine. But it's clear the body has remarkable healing powers that we are only just beginning to discover and understand, which raises more questions.

If we can, through modern science and engineering, learn to build new legs, reprogram and change the characteristics of the ones we have—even grow new ones—what else might we be capable of creating? And what other untapped powers for regeneration, resilience, and transcendence lie within us, in other areas of the body?

What might we do for people limited in other ways—people whose deficits, for instance, are not in the area of movement, but who lose their ability to sense the world around them instead?

PART II

SENSING

CHAPTER 4

THE WOMAN WHO CAN SEE WITH HER EARS

NEUROPLASTICITY AND LEARNING PILLS

The last thing twenty-one-year-old Pat Fletcher saw before the explosion was the chemical-filled steel tank beside her suddenly ballooning outward. With alarm she realized the plastic hose in her hand had grown unusually hot. Then the world flashed blindingly bright and turned a brilliant blue, the color of the flames engulfing her body.

When she awoke, Pat thought she might be dreaming. The world around her was featureless and dark, as though she were lost in a gray, smoky fog. The sedatives and painkillers had something to do with it, as did the fact that her face was swathed in thick bandages. But soon a solemn doctor arrived at her bedside. And Pat learned there was something more. She had been in an industrial accident caused by a reaction between two volatile chemicals at the grenade factory where she worked. One of her eyeballs was gone; the other eye remained, but was permanently shut. Pat was lucky to be alive, the doctor told her. But there was no hope she would ever see again.

It would take nearly three decades, but, in a way, the doctor was wrong. Twenty-five years later, the extroverted, gray-haired resident of Buffalo, New York, was surfing the Internet using a program that converts the text on the screen into speech when she stumbled across a computer program designed by a Dutch engineer. He claimed his program, which he called "vOICe," could convert the pixels in images into sounds and allow blind individuals to "see" the world around them. Pat was dubious. She even chuckled when she played a sample "soundscape," a pastiche of scores of different tones at different volumes and pitches emitted simultaneously. It seemed absurd. An unintelligible jumble of noise.

Then Pat cranked up a "picture" of a long, gated barn fence through a pair of stereo speakers in her study, and it just about took her breath away. Something was happening in her mind's eye, something that felt fundamentally different than simply "hearing" the sounds.

"I turned around and I could almost just see the fence going all the way across my study and I said, 'Oh Lord, what is this?'" Pat recalls. "I just started getting chills up my back."

What made the feeling so unbelievable was that she could tell the sound was *out there*—beyond the reach of her cane bumping into something, beyond the pull of the leash on her hand as her dog guided her forward—beyond her touch. From the dynamic cacophony of sound, somehow, Pat had a sense of the fence's dimensions, of its shape and where there were gaps between the slats. The world of the blind has often been described as profoundly claustrophobic, because all that is knowable and perceivable about the shapes and the objects that surround them ends abruptly at the end of one's fingertips. But Pat's world had just expanded.

How could sound do this? she marveled.

"It felt like the image was real," she says. "It's a fence—'see there is a gate—and there's a blackness there, like the gate is open . . .' It was a shock. It just felt like you could walk along it and that really, really shook me up."

Pat went to the store and purchased the smallest webcam she could find, attached it to a baseball cap, and then hooked it up to a laptop computer. Then she turned it all on, walked out into her hallway, and looked around.

"That almost took me to my knees," she says. "I could tell that there was a wall, and I walked up to the plastic window blinds and touched them and I just couldn't believe it. You just forget what the world looks like."

Pat soon discovered she could see patterns on drinking cups she'd been blind to for years. She got lost in the decorative wallpaper in her dentist's waiting room. She could see leaves moving on the trees. She could see faces, though they remained blurry. Pat sent away for a pair of spyglasses with a camera hidden behind a tiny hole at eye level and upgraded her rig. She began to use the device every day. Soon she carried her cane only so she would have it on hand in case of a technical malfunction.

And then one afternoon four years later, something even more amazing happened. Up until that day, when she gazed into rooms, or looked around, it was almost as if she were looking at a flat, two-dimensional photograph. She could see there was a couch in the living room, or the shape of a tree against the sky, but she had no sense of depth. But that day Pat was standing at the sink washing dishes, when she stepped back to dry her hands on a towel and looked down. The sink had always appeared to her as a simple square. But with her new device on, Pat realized suddenly she had regained depth perception.

Pat Fletcher was looking *into* the sink.

Pat Fletcher's experience sounds impossible, or at the very least like some elaborate kind of mind trick. Maybe *she's* convinced. But it can't be real—after all, it flies in the face of conventional scientific theory. It flies in the face of conventional wisdom. How can you "see" with your ears? How could the brain suddenly rediscover the capacity for depth perception four years out, seemingly as suddenly as if someone had flipped a light switch?

Yet Pat Fletcher's claims have been verified by some of the world's leading scientists. Several years ago, the intrepid fifty-eight-year-old technological adventurer, wearing her jerry-rigged device, arrived in Boston for testing at Harvard Medical School. Pat lay down on a large table, which slid her into the cramped tube of an MRI machine capable of tracking the amount of oxygen being used by different parts of her brain. The doctors instructed her to listen to her soundscapes.

Pat Fletcher still had no eyeballs with which to gaze upon the world. Yet somehow, when she listened to her "soundscapes," the areas of the brain associated with visual processing in the sighted—the areas of the brain normally activated when we point our eyeballs at an object in space—sprang to life. Meanwhile, when Pat heard normal sounds, when, for instance, a researcher jingled his keys nearby, Pat's auditory cortex continued to light up like normal. Her brain somehow was able to distinguish between normal sounds and her soundscapes and route the latter to the correct area of the brain for processing vision—even when these sounds entered her ears simultaneously.

A series of additional experiments appeared to confirm it. Pat Fletcher, blind for more than thirty years, was in some sense seeing (while sometimes simultaneously hearing) with her ears. Her brain had rewired itself.

Throughout the centuries, we've collectively made profound scientific efforts to restore our senses when they weaken or fail, from the seemingly mundane (the hearing aid, nasal spray) to the more advanced (cochlear implants and LASIK surgery). After all, our five senses—vision, hearing, touch, smell, and taste—are our portals to the world, our connection to each other.

For decades, scientists have focused on enhancing or somehow repairing the *external* machinery—our eyes, ears, nose, taste buds—in other words, the physical parts of our bodies that directly interface with the world around us and suck up the sensory data.

But recent advances are offering a whole new array of devices, like Pat's soundscape machine, that are fundamentally transforming the way we think about how the brain processes the information we receive from the environment. They suggest that for decades scientists have perhaps missed the point.

"We see with the brain, not with the eyes," the late neuroscientist and pioneer in sensory substitution Paul Bach-y-Rita famously declared. "You can lose your retina, but you do not lose the ability to see as long as your brain is intact."

The brain is perhaps the world's most sophisticated pattern recognition machine. And given time to reorient itself, it is far better able to make sense of foreign and new patterns of sensory inputs than previously realized. To restore something very much like sight, scientists are discovering, it is not necessary to replicate the exact patterns of electrical pulses that the human eye sends the brain. Nor to give a deaf person hearing is it necessary to figure out the exact timing and patterns of pulses the human cochlea uses to encode sound waves that enter the inner ear. We don't necessarily need to fix the broken *external* parts or replace them with exact replicas to regain lost function.

To restore "sight" or sound, all engineers need to do is create a device that can translate sensory information into signals that can be conveyed to the brain in a consistent manner. Given sufficient practice, the connections and pathways in the brain begin to rewire themselves and in so doing can learn to decode sensory information delivered through a wide array of means. Pat Fletcher's device is just one example of a new class of technologies taking advantage of the remarkable plasticity of the human brain—plasticity that until recently most believed largely disappeared after critical periods of childhood.

In the summer of 2002, Pat Fletcher packed a couple of carry-on bags—one for clothing, one for her soundscape equipment—and hopped into a taxi to the Buffalo airport. She'd been using the vOICe for a few years now. She didn't have a guide dog anymore—she didn't believe in keeping a dog on a leash now that she no longer needed it. She had her cane and she had her soundscapes, and she was ready to go.

The terrorist attacks of 9/11 still loomed large in the United States, and at the airport metal detector her equipment bag set off alarm bells that delayed her in a thicket of agitated security personnel for more than half an hour.

"To tell someone I'm blind—they could see that," she says. "But then to see all those wires taped to batteries in my black equipment bag made them really nervous. I turned on my computer and showed them how it worked and they were reassured."

Pat was headed to a conference in Arizona for devotees of "consciousness studies" and specifically of a new field called "sensory substitution."

Once there, she learned about a device called the "tongue camera" that transforms images into electrical pulses on the tongue, which amazingly seemed to work. She heard about the possibility of creating a machine to convert sounds to electrical pulses on the skin and help the profoundly deaf hear again. But for the other conference participants few technologies could actually compete with Pat Fletcher herself. She was in her element, and she was a star.

At the conference, Pat took the stage with Peter Meijer, the soft-spoken Dutch engineer who had created the software she used for her device. To her, it was like meeting Henry Ford or Thomas Edison. It was special for him, too. Meijer, a slim wisp of a man, with a broad forehead, a mop of brown hair, soft, brown eyes, and a thin, modest smile, had spent almost a decade tinkering on his spare time in his living room before coming up with something that worked. Yet Pat was the first blind person he had actually met face-to-face who had worn the device so regularly. He was as delighted as she was by the results.

In between sessions, they took walks around the conference center campus together and just talked. Peter asked her lots of questions—he wanted to know how to improve the device. He wanted to know what it felt like for her. After a couple of days, they were like old friends. And when the organizers took the conference participants to a museum out in the desert, the two slipped away and stood outside on a hot, dusty plain gazing out at the horizon.

Pat had always loved nature. One of the hardest parts of losing her sight had been the idea that she wouldn't be able to hike anymore.

"I could barely make it to the bathroom; how in the heck was I going to make it up a trail?" she remembers thinking sadly in those first devastating days. "How was I going to look out across the ocean and watch the storm clouds battling themselves? That's gone, you know? How could I get in a boat and go fishing by myself? It just wasn't going to happen."

Pat hadn't "seen" much of nature since the accident. Now, as they stood in that desert, Peter told Pat to look up. She could see a streak in the sky and asked him what it was.

"A smoke trail from a jet," he told her.

"But what are those things way out there?'" Pat asked. She could see a bunch of triangles with points, at different levels, standing out from the whiteness of the sand. "You can see those?!" Meijer asked, incredulous that she could both see something so far away and perceive its distance.

"Uh-huh, yes."

"Those are mountains," he told her.

The next thing Pat knew, the tears were streaming down her cheeks. She was so excited, she had started to cry. Pat saw things better up close—she could see the shapes of the big cactuses, their arms, and even the edges and gullies rippling up and down the plant (though she couldn't see the needles). In the distance the mountains appeared as simple triangles, lined up against one another at varying heights. But it was enough.

"I love mountains," she says. "They're one of my favorite, favorite things in the world, and here I was able to see a mountain range again. I get emotional when I see things I never expected to see again. I just couldn't believe it."

It was perhaps only then that Meijer, a polite, sweet, but famously modest engineer from Eindhoven, fully appreciated the powerful emotional impact of his creation, which he called the "vOICe" because the three letters in the middle could be read as "Oh I see." The two stood in silence and allowed the beauty of the moment to sink in.

Meijer hadn't expected to truly transform people's lives. He originally came up with the idea for a machine to convert sight to sound while still a graduate student studying physics, because he was looking for a way to play around with new computer technologies. He wasn't thinking about neuroscience. He just wanted to build something useful. That's how he got it in his head that it might be cool to build what he called a "reverse spectrograph" to help somebody like Pat use hearing to judge images.

Spectrograms are charts that visually represent sounds. The horizontal x-axis on a spectrogram chart represents the passage of time. The vertical y-axis represents frequency, which humans perceive as pitch. If you were to run a fingertip across a spectrogram representing one tone, from left to right, you could trace the peaks and valleys of connected dots unfolding above this horizontal x-axis and get a pretty good sense of the rising and falling pitches of the sounds it portrays over time. The farther you move from right to left, the more time has passed. The higher up the dots, the higher the pitch. The amplitude, or volume, is represented by different shades of gray—the brighter the shade, the louder the sound. Most spectrograms portray multiple dots stacked on top of one another at any given point in time, to represent all the tones played together at a single moment. They are commonly used in speech analysis—you might have seen one in a spy thriller, where the bad guys are monitoring the phone lines for the vocal signature of the hero on the run.

Meijer's idea was to create a machine, a decoder, that could do the reverse—turn the visual dots, or pixels in an image, back into sounds. His first prototype was bulky. And the jerry-rigged device Pat took off her head to show him years later at that conference worked better than he could have imagined—given the technology of the time—when he started out. First, the tiny spy camera in between the lenses of Pat's sunglasses gathered footage, which in its digital form ran through a computer program, a Rosetta stone of sorts, a master decoder. Meijer's algorithm then translated each pixel into the appropriate tone.

The higher up a pixel was in a column, the higher up the corresponding tone emitted by Meijer's devices. The brightness of the pixel corresponded to loudness of the sound. The time or the pace of "stereo panning" and the resulting sounds represented the changing horizontal characteristic of the image as the camera captured it. Basically, the pastiche of blended tones went up and down in a series of sound waves as the system scanned across the face of an image, encoding its contours in the oscillations of pitch. The result was like a turbocharged inkjet printer spitting out dots to form a picture, only the vOICe moved far faster and spit out oscillating sound waves instead of ink. The key to the device was its consistency. Specific shapes produced specific sounds. Specific patterns of pitches encoded specific contours. Over time, somehow, it seemed, the brain could learn to associate these patterns of pitches with corresponding contours in the physical world, and the specific sounds with shapes.

The device also worked because the human ear is capable of distinguishing a surprising multiplicity of tones at the same time—30 to more than 100, depending on the image, Meijer would later find out. At each interval of time, Meijer's device played an entire column of pixels, represented by multiple tones of various pitches and volumes simultaneously. By moving across the image from left to right extremely rapidly, Meijer was able to convey a tremendous amount of information in a very brief period of time. Remarkably the brain was capable not just of discerning this huge variety of tones, changing at an extremely rapid clip, but of almost instantaneously analyzing them and comparing them to previously learned patterns of sounds in a way that allowed Pat to immediately comprehend that what she was "looking" at was a gate or a window shade, or a pattern on a coffee cup.

For Meijer, it was a long-term project, one he toiled on for years in his one-room apartment at nights and on the weekends. By the time he finally finished, it was the early 1990s, and Meijer was working in the R&D arm of Philips, the Dutch technology behemoth. Though Meijer's specialty was building simulations for new kinds of computer chips, he took his invention to his superiors, who patented it and encouraged him to publish a paper on the device.

"I had a lot of interest, questions from around the world, people who wanted reprints; it was amazing, hundreds of requests for reprints," Meijer recalls. "But after a while, we had one luggable prototype that you cannot do too much with. We can give demos, but you cannot effectively train someone."

In fact, when Meijer contacted organizations for the blind, he was met by skepticism and indifference. Despite the initial interest from intellectual quarters, when he came to those who could really use the device, it seemed, nobody was sure what to make of him. So Meijer put the device on the Internet and encouraged blind individuals and university researchers to pull the software down and experiment with it themselves.

They waited, in other words, for someone exactly like Pat Fletcher to come along and take the device to the next level.

Meijer and his colleagues weren't the only ones waiting for a Pat Fletcher to come along. At Boston's Beth Israel Deaconess Medical Center, a dapper Spanish-born Harvard neuroscientist named Alvaro Pascual-Leone had access to millions of dollars in brain-scanning equipment and an entire program aimed at studying the plasticity of the human brain that was set up to examine people just like Fletcher—people, in other words, whose experience defied conventional wisdom; people who could teach us new things about how the brain works.

The problem with studying such unusual subjects, however, is that they are never easy to find.

So, when Pascual-Leone first read about Meijer's device in a scientific journal, he made it a point to take a side trip to the Netherlands the next time he was on a holiday visit in his native Spain. In August 2001, Meijer set up a demonstration for the Harvard neuroscientist in his modest home workshop, equipping Pascual-Leone with a blindfold and headphones and letting him experience the soundscapes for himself.

"It was absolutely spectacular because very, very simple things I could sort of get, at least some sense of it," Pascual-Leone says. "But the more complex images and situations and things, I had no idea what in the world I was listening to. It was profoundly disorienting and it made no sense whatsoever to me."

Then Meijer mentioned that there was a cheerful spitfire of a woman in Buffalo who had found his program while performing an Internet search a few years back and she had actually learned how to use it in the real world. This woman, Meijer told Pascual-Leone, claimed she could "see" with her ears. Would Pascual-Leone like to speak with her?

"'You gotta be kidding me!'" Pascual-Leone remembers answering. "I was blown away."

For decades, debate had raged about the capacity of the adult human brain for change. Few doubted the brain could radically rewire itself in early development. But it was widely believed that this period was finite—that after a brief "critical period" in immature humans and animals, the connections of the brain hardened, like clay in a kiln, and became fixed.

Pascual-Leone, however, was part of a small group of renegade scientists who believed that this view was far too simplistic. And though Pat Fletcher's case sounded extreme, Pascual-Leone had seen enough not to dismiss it out of hand. If what Meijer said was true, this was a wild and potent example that the brain could indeed rewire—a paradigm-busting example.

One afternoon, the phone rang in Pat Fletcher's Buffalo home. Forever after, she'd get a kick out of telling people about it. "Harvard University wanted me to come for a visit!" she'd crow. "Can you believe it?"

To understand why Pat's claims sounded so remarkable, it helps to first understand the work that convinced many neuroscientists such a thing was impossible.

The seminal evidence came from a series of groundbreaking experiments performed on the visual cortex of cats and monkeys by David Hubel and Torsten Wiesel in the 1960s and 1970s. These experiments vastly expanded our understanding of how the brain allows us to perceive the world around us, and would eventually win them the Nobel Prize. And if you'd read about them and studied them closely, as Pascual-Leone and every single neuroscientist of his generation undoubtedly had, it was not so easy to explain Pat Fletcher's remarkable abilities.

David Hubel, a Canadian scientist, met the Swedish Torsten Wiesel in the late 1950s at Johns Hopkins. The two young researchers were only a decade or so older than Pat Fletcher had been when she'd lost her eyesight. They were young postdocs, at the start of their careers. And they had grand ambitions.

Together they set up in a cramped and dingy, windowless room in the basement of the university's renowned eye institute and set out to solve a mystery that had long flummoxed scientists: What exactly happens in the brain when we "see" objects and shapes? It was an auspicious time to ask the question. Brain scientists had only just begun to use a revolutionary new technique called "single-unit recording." While tinkering with a lathe and different materials, Hubel himself, in fact, had come up with a new way to make tungsten-wire electrodes so effective at recording from the brain that many other scientists were embracing his technique. The magic of single-unit recording was that it allowed scientists for the first time to monitor in real time the activities of the most basic unit of the brain, individual nerve cells, called "neurons."

Every one of us has about 100 billion neurons, separated from one another by tiny structures known as "synapses." Neurons talk to each other by passing electrochemical signals across these synapses. The branches of the neuron that lead from the cell's nucleus to the synapse and transmit the signals are called "axons." The branches that receive signals at the synapse and carry them to the cell body are called "dendrites." If strong enough, the signals one neuron's axons pass on to a second neuron's dendrites will cause that second neuron to fire electrical pulses. When a neuron fires, it passes electrochemical messages of its own through its axons on to the neurons it is connected to, which in turn makes those neurons more likely to fire. Other kinds of synapses "inhibit" connected neurons from firing.

The firing of neurons in concert or in sequence has often been described as a "symphony," a beautiful, coordinated conglomeration of many instruments playing together, producing a whole greater than its parts. (I first saw the metaphor used in a book by Duke University's Miguel Nicolelis.) It's this symphony that allows us to think, to feel, and to move.

And to see. Sitting in that windowless Baltimore basement, Hubel and Wiesel hoped to listen to the symphony, and its individual parts, as few before them had. To do so, they anesthetized a cat and inserted tiny pin-shaped microelectrodes directly into its gray matter. The microelectrodes picked up the sound of the cat's individual neurons firing and transmitted them out of the brain to an amplifier, which then filled the room with a distinctive clicking sound every time a neuron fired. These signals could also be converted into visual representations and flashed on a screen or graphed, which allowed the young scientists to examine the frequency and duration of each spike.

The territory Hubel and Wiesel planned to explore was in the outermost layer of neurons, just below the skull and a thin sheen of protective coating, in areas known to comprise the primary visual processing centers of the brain. They were located in the back of the head, in the cerebral cortex, a piece of tissue 2 to 4 millimeters thick, rich in neurons, and essential not just to our ability to move, sense, and react to our environment, but to almost every type of higher-level processing that separates us from our reptilian ancestors.

Neuroscientists had been successfully homing in on the large-scale organization of the cerebral cortex for a hundred years by, among other techniques, studying lesions in stroke victims and creating new ones in animals. (If a stroke killed brain cells in a particular area of the cortex, and their death was associated with loss of a specific function, such as speech, one could infer the brain cells that had been killed played a role in the ability to carry out that function.)

But the precise details on the micro level, the way individual neurons worked together—the actual functional organization of the brain's billion-strong army of worker cells—remained largely a mystery. It was virgin territory, wide open to the two ambitious young scientists.

For Hubel and Wiesel, the days back then were long, the work somctimes frustrating. Often they worked until they were so tired Wiesel would begin addressing Hubel in Swedish. (That's when they knew it was time to call it a night.) At least once, Hubel returned home just as his family was sitting down for breakfast. But the efforts began to pay off about a month in. The pair had set up a cat in a head mount sturdy enough to survive an earthquake. A jumble of wires led to the electrodes connected to neurons held in place by headgear, recording how specific neurons in the cat's visual cortex responded as the scientists used a device to project different patterns and shapes directly onto the surface of the anesthetized feline's retina.

The goal was to find a single stimulus that could cause one, specific neuron to fire in the visual cortex. Hubel and Wiesel tried projecting dark spots on light backgrounds. They tried bright spots on dark backgrounds. They varied sizes. Eventually they even waved their arms, danced around in front of the cat, and, to lighten the somber mood, began showing it pictures of sexy women in magazine advertisements. Nothing. The neuron lay dormant.

Then one day about four hours in, they were putting in a new glass slide of a black dot when the neuron "went off like a machine gun," Hubel would later recall. The sound of the neuron firing had nothing to do with the dot itself. As they had inserted the slide, its edge had cast a faint but sharp shadow onto the cat's retina—a shadow of a straight dark line on a background. The neuron they were studying, they realized, was primed to respond most powerfully to that line.

Some individual neurons, they would soon conclude, fired with greater intensity in response to lines at *specific angles,* while others appeared primed to fire at angled lines *moving* in a specific direction. In other words, specific neurons of the brain were assigned to respond to and represent specific stimuli from outside the body. They had "receptive fields," and in many cases if a stimuli was in the center of that field, the neuron fired full bore. Stimuli on the edge of the receptive field caused slower firing. Anything outside it left the neuron dormant.

It is the firing of these individual neurons in concert, Hubel and Wiesel had discovered, that helps us to build complex images in our mind's eye. These neurons are arranged in columns, and visual analysis takes place in a well-ordered sequence as electrical signals pass from one nerve cell to the next, with every nerve cell largely responsible for particular kinds of details in a picture pattern.

The signal message that the eye sends to the brain has long been a "secret code to which only the brain possessses the key and can interpret the message," Professor David Ottoson of the Karolinska Institute would later say when he presented the Nobel Prize in 1981. "Hubel and Wiesel have succeeded in breaking the code."

Hubel and Wiesel still wanted to understand how these visual cells developed. How did a neuron come to respond to a diagonal line or a vertical edge? Why were some sensitive to movement? And how did these neurons come to work together to form an image—to become part of a larger visual processing circuit?

Hubel and Wiesel suspected experience played a key role. Children born with defects in the lens of an eye that blocked light, a cataract, often suffered permanent visual damage, even after the cataract was surgically removed. Yet there was no such damage in elderly cataract patients. How to explain this discrepancy?

To create a similar circumstance in kittens, Hubel and Wiesel sewed one eyelid shut and allowed the second eye to develop normally. Then they repeated the experiment with mature cats. In the adult cats, unsewing the sutured eye restored vision. But in kittens, the formerly sutured eye remained permanently blind even after they had uncovered it. Hubel and Wiesel had what seemed irrefutable proof that the brain has what they called "critical periods" when it develops and can be programmed. It was an electrifying experiment, but it raised far more questions than it answered. How did these critical periods work? Could they be reversed? What was the biochemical basis for this change in the brain?

In recent years, neuroscientists have been able to watch the neural circuits form in brains of young animals in real time in response to stimuli, and have answered some of these questions. One of the most elegant such experiments was conducted by Hollis Cline, a neuroscientist at the Scripps Research Institute who served as president of the Society for Neuroscience in 2015–16. In the mid-1990s, she used a technique called "two-photon microscopy" to peer into the brain of a tadpole and witness at an unprecedented level of resolution the way the neurons formed their initial connections in the developing brain.

What she saw was a picture far more dynamic and graceful than anything ever before detailed. In the brain of the tadpole, the branching projections from different neurons were constantly growing and retracting, reaching out toward one another like long, delicate fingers seeking contact. Most of the time, the contact between the branches of different neurons was as fleeting as a quick bump—they would touch, and then quickly retract, bouncing off one another and continuing on their way to connect with others.

But every once in a while something happened that caused the two branches to stick together in a semipermanent embrace. This magical connection only occurred if both of the branches were attached to cell bodies that happened to be firing at the moment of contact. Cline had captured the moment when synapses, the microscopic connections between two cells, were born.

The phenomenon Cline was witnessing was made possible by a principle that scientists had long suspected but have only begun to prove in recent decades. In 1949, a Canadian psychologist named Donald Hebb suggested that the brain was essentially a powerful coincidence detector, and that the physical rules governing how connections between neurons form and strengthen are designed to reflect and record these coincidences. Thus when two neurons fire in close succession, something occurs in the brain that strengthens their physical connections to one another—something that makes it easier for one neuron to activate another in the future. On the other hand, when two neurons fire separately, their connections weaken. This is often called "Hebbian learning."

It was a principal perhaps most articulately summed up by Carla Shatz, a young researcher in the lab of Torsten Wiesel and David Hubel in the early 1970s, who is now a leading researcher on "plasticity" at Stanford University.

"Cells that fire together, wire together," wrote Shatz, who actually helped prove this was so by measuring increases in the voltage passed between connected neurons. "Cells that fire out of synch, lose their link."

In the human fetus, many of the initial connections between neurons form in exactly this way. Spontaneous electrical impulses pulse through the brain in random patterns, and the young neurons dance, bump, explore, and promiscuously hook up. Most neurons, in this initial stage, will form far more connections than they need and will maintain. Later, as development progresses, these connections will either strengthen each time they happen to fire in close sequence again, or the connections will gradually weaken. Eventually extraneous connections will be clipped away like stray branches in a hedge, in a process scientists call "pruning."

Over time, the connections between individual neurons, coupled with the pruning of irrelevant connections, form the brain's circuits, the tightly connected, superefficient infrastructure of our visual or auditory systems, for instance.

Hubel and Wiesel's experiments with the kitten and the sutured eye suggested that, in some areas of the brain, this mechanism of circuit formation was only in play for a finite period of time, during so-called critical periods. After these critical periods closed, it appeared, it was too late to change. The clay had hardened. The circuits had formed. The pathways in the brain were set in place.

Perhaps even more remarkable, Hubel and Wiesel also discovered that the cortical real estate in the kittens normally used by the blind eye did not go to waste. Instead, the connections conveying sensory information from the dominant eye branched out, expanded into the unused space, and took it over. Ever efficient, the brain apparently operated on a principle of "use it or lose it."

It's difficult to overestimate the impact these findings would have in the decades that followed on brain science and the way researchers looked at development. The idea that the brain has critical periods in which it is highly plastic after birth—and that these periods slammed shut—had implications far beyond just the development of vision.

Within a few years, the scientific establishment had not only embraced Hubel and Wiesel's findings on critical periods in the visual system. Many had taken the idea much further, concluding that most, if not all, plasticity mechanisms in the cerebral cortex disappeared with age. After all, it seemed to explain a lot. Why, for instance, it was so much harder to learn a second language with a flawless pronunciation as an adult; how as we grow older we become "set in our ways"; why children are so much more driven to explore, learn, and question than adults are.

A number of untested assumptions soon took on the power of conventional wisdom. Many in the medical establishment believed stroke victims could never recover function—since strokes laid waste to vast areas of neural territory, and the brain was believed incapable of rewiring itself in adulthood to go around the dead areas. Educators argued that those with dyslexia and other learning disabilities could never fully overcome them—they were just wired that way. Certainly, the idea that someone like Pat Fletcher who lost her eyesight in adulthood could eventually learn to "see" with her ears would have seemed absurd.

But almost as soon as these experiments came on line, evidence began to emerge that there might be some exceptions—that, in fact, things might be a little more complicated than they first appeared. It would take decades to overcome the entrenched dogma, a process that only seems to have succeeded since the turn of the millennium. But it is now widely accepted that stroke victims *can* recover function. Dyslexic children *can* learn to read. And, perhaps, Pat Fletcher can see with her ears. Because though Hubel and Wiesel got much of it right, the brain *does* retain a remarkable amount of plasticity into adulthood.

To change the brain, we just needed a better understanding of how it works.

Perhaps the most potent early hint that things weren't as simple as they at first appeared came from a device that, like Pat Fletcher's soundscape machine, used the machinery of the ear as an entry point to sensory-processing areas of the brain. That device is the cochlear implant.

One of the pioneers of the implant, a man whom many would later come to consider the "father of neuroplasticity"—and whose work would have a profound impact on Pat's new friend Pascal Alvaro-Leone—got his start, ironically, just a few hundred feet away from the hallowed halls where Hubel and Wiesel had cut their teeth, in the neuroscience buildings of Johns Hopkins.

His name was Michael Merzenich, and in the beginning he wasn't looking to overturn convention. Far from it. When Merzenich arrived to pursue his Ph.D. in Baltimore, about five years after Hubel and Wiesel had departed their nearby office for Harvard University, he had every intention of pursuing a conventional path.

Later, though, Merzenich had a series of experiences that would profoundly affect his worldview. While doing postdoctoral work at the University of Wisconsin–Madison, Merzenich along with his collaborators were investigating a bizarre phenomenon that sometimes occurs after damage to the large nerves that carry signals between the brain and the skin of the body. Unlike the central nervous system, "peripheral" nerves—such as those that carry signals from the skin of the hands—are capable of regenerating themselves if they are severed. When the nerve from a hand is cut, however, sometimes when it grows back the signals get scrambled; if you touch a middle finger that has suffered nerve damage and repaired itself, for instance, you might feel the sensation forever after on your thumb.

To better understand what was going on, Merzenich and his team used electrodes to record the activity of individual neurons, a technique Merzenich had mastered at Johns Hopkins, to map the area of the cortex that processes touch in the brain of normal adolescent monkeys. Then they severed the peripheral nerve carrying signals from the monkey's hand to this part of the brain, the somatosensory cortex. The site of the cut was in a bundle of fibers leading from three fingers and the palm of the hand in a bundle of nerves that traveled up the spinal column to the brain.

Next, Merzenich and the team sewed this bundle of fibers back together so that they were almost, but not quite, touching. This allowed the nerves to regenerate and reconnect but in a random order. Merzenich and his team expected they could then understand the process that created the distortions in sensory feeling that occurred after the nerves from that part of the hand had been crossed.

But when Merzenich and his team went back and remapped the same area of the brain seven months later, they were amazed at what they found. The nerves had indeed been jumbled—but the brain had created a new order. It had completely *remapped* itself to account for the crossed nerves, resulting in a patchwork of signals that had not been there before. And it had done so at an age long past the end of what many assumed would be a "critical period." The brain had done something that most would have assumed was impossible.

After that, Merzenich says, "I knew the brain was plastic and it was changing itself."

Merzenich didn't understand just how plastic it could be—and just how profound the implications—until a couple of years later. After finishing his graduate work, Merzenich took a job at the University of California, San Francisco (UCSF), joining the department of otolaryngology and physiology as an assistant professor and focusing on the ear. Soon after arriving, he met a surgeon named Robin Michelson. Michelson wanted to build a device that could help deaf people hear again, using an approach others were also pursuing at the time that aimed to build what would become known as "cochlear implants." Michelson asked Merzenich if he might be willing to help.

"He was a bold adventurer," Merzenich would later recall, "who had gotten an engineer in L.A. to make a device and implant it in some patients, but he didn't have the knowledge to refine it."

Merzenich thought Michelson's approach showed promise, so he signed on. The human ear is designed to detect, with tiny hairlike structures, the vibrations that constitute sound, convert them into electrical impulses, and send them to the brain for processing via the auditory nerve. It's a process that takes place in a seashell-shaped chamber of bone in the ear known as the "cochlea" (the Greek word for snail), a gateway into the auditory processing areas of the brain.

Rather than simply amplify sounds, the traditional approach to hearing aids, cochlear implants use long, threadlike arrays of stimulating electrodes to directly shock the auditory nerve in patterns that mimic the electrical pulses generated in the auditory nerve by fully functional ears. In the beginning, Merzenich assumed that to make a cochlear implant he would have to come pretty close to matching the patterns of electrical pulses generated by the movements of the hairlike fibers in a functional human ear.

But this was not so easy to do. The device had to be remarkably sturdy, with electronics capable of surviving decades. Yet it had to be safe enough for a surgeon to be able to implant it in a fragile human hearing organ. The largest challenge was simply that the exact pattern of electrical pulses created in the cochlea and sent through the auditory nerve to convey sounds to the brain turned out to be far too complex, far too detailed, to capture with existing technologies. The grand ambitions of Merzenich and his collaborators were quickly frustrated—they could only crudely replicate the "elegant, refined pattern of the intact inner ear."

"It's like playing the piano with your forearms," Merzenich would later tell me. "You can't really control the details. It codes the information in a relatively crude way."

When Merzenich and his collaborators implanted the first models in patients, and the patients began arriving at Merzenich's office for follow-up visits, they quickly confirmed the researchers' worst fears.

"They said it was 'crap,'" Merzenich recalls. "The sounds they were hearing were totally muffled and muddled and messy, and uninterpretable. Just garbage."

Merzenich and his team tried a number of approaches to

improve the device. But no training regime, experiment, or tweak really seemed to make a substantial difference.

Yet the experiment plodded on. It had to. Merzenich's team had already implanted the devices at great effort and cost. And it wasn't as if the patients had anything better going. They were profoundly deaf, condemned to an invisible cone of silence in a world where they were surrounded by people talking and conversing and joking and laughing. Excluded. They weren't eager to quit, either. So the devices stayed in.

It's a good thing they did, because after a few months, something amazing began to happen.

Merzenich remembers the first patients, formerly glum and not particularly pleased with the device, arriving for their check-ins suddenly bursting with enthusiasm.

"I'm beginning to understand everything now!" they told Merzenich. "Wow!"

Something was happening; the tests confirmed it.

"Over a two- or three-week period, they have this amazing sort of clarification of what they're hearing," Merzenich recalls. "Suddenly they can demonstrate that they can understand at a relatively high level. It was shocking."

There wasn't anything different about the devices. It was, Merzenich realized, the brains of his patients that had changed. At the time, in his laboratory Merzenich had been pushing forward with his exploration of neuroplasticity in monkeys in parallel to the cochlear implant project. But even he was astonished.

"The brain could take the information of these crude signals and turn it into a new form of representative speech," Merzenich says. "These devices worked better than we had imagined. I just didn't think the brain could change on that scale."

What's more, Merzenich soon learned that researchers in competing groups creating cochlear implants were getting similar results—and they were using entirely different coding schemes and patterns of electrical activity.

What seemed to matter was not the details of the patterns of the signals going to the brain themselves, but their consistency. The brain is essentially an elaborate pattern-recognition machine. It is dynamic and ever changing, and capable of learning to associate specific pulses of electrical stimuli with specific sounds and words meant to represent ideas in the outside world. It can even do so when those pulses and combinations are far cruder and totally different than those produced by the native-born machinery of the ear. Hebb's rule of associative learning, the idea that neurons that fire together will eventually wire together, was far more powerful, far more robust and enduring, than even Merzenich could have guessed.

"You put a new front end on the auditory system, wait six months, and the brain didn't care," he says. "It was amazing."

Pat Fletcher arrived in Boston for her first tests in the laboratory of Alvaro Pascual-Leone on a hot summer day in 2006. Pascual-Leone's team put her up in a bed-and-breakfast near Beth Israel Deaconess Medical Center, the location of the labs where they intended to conduct the tests.

They'd also flown in another blind test subject named Adam Shaible, along with his wife, Denise. Pat and Adam had corresponded over an e-mail message board set up by Meijer to bring vOICe users together, but it was the first time they had met face-to-face. Every morning before the testing, Pat, Denise, and Adam convened at the communal dining room table in their cozy inn and swapped stories.

Adam told Pat the joy of "seeing" his wife's face and hair for the first time, how much he loved just to look at her. Adam lived in Florida, not far from the site of Pat's accident so many years before. And Pat could almost picture it herself when Adam described the first time he stood on the shore and watched with wonder as a majestic sloop glided across the pristine waters of the bay. Pat remembered the sailboats, the way their canvas sails could dance with the wind, from her time as a sighted person. She'd loved them.

Pat nodded knowingly when Adam told her about spotting ripples on potato chips for the first time. And she had to laugh when Adam described how weird and surprising it had been to see steam rise up off the top of a coffee cup.

"I don't think about that, because I am used to seeing the steam off the coffee!" she said. "But imagine if you'd never seen it before."

But something nagged at Adam, tormented him. Like Pat, people often told Adam, "There's no way you can see." Pat knew what it felt like to see something. She remembered. And she had no doubt that what she was experiencing was indeed sight. It was different for Adam. Unlike Pat, Adam was born blind. How could he be sure it was real—that what he was experiencing was actually this thing called "sight" that he had been hearing about all his life?

At the lab, Pascual-Leone and his team had designed a series of elaborate experiments. And they assured Adam that by the end, they would be able to at least partially answer his question. In fact, Pascual-Leone suspected he already knew what the answer would be, because in some ways the experiments he would perform on Pat and Adam were simply the next step—a confirmation of a substantial and growing body of research.

A native of Valencia, Spain, Pascual-Leone earned an M.D. and Ph.D. in Germany, and studied neurology at the University of Minnesota before moving on to the National Institutes of Health (NIH) in Bethesda.

Even before he arrived at the NIH, Pascual-Leone had been following the work of Mike Merzenich with deep interest. The rumpled Californian had been busy since that first monkey experiment and his initial cochlear implant experiences. He had, in fact, gone on to conduct a series of experiments with monkeys that Pascual-Leone was itching to take to the next step—in humans. These experiments had profoundly affected Pascual-Leone's worldview.

While on sabbatical from UCSF, Merzenich had teamed up with Jon Kaas, of Vanderbilt University, and together they had designed a radical experiment to find out just how "hardwired" the brain is after critical periods had closed. To do so, they once again used microelectrodes to extensively map the area of the somatosensory cortex of a monkey that appeared to register nerve inputs from different areas of one hand.

Then they cut the peripheral nerve connecting the palm of the hand to the pathways that carried signals to the brain. This time they did not sew it back together. Like cutting a telephone wire, all communication between that part of the hand and that part of the brain devoted to processing its inputs went dead.

After several months, Merzenich and Kaas went back in and remapped the brain to see if anything had changed. Hubel and Wiesel had shown that unused cortical real estate in the visual centers of a kitten's brain could rewire itself to perform a different function when one eye went unused—but only during critical periods of development. The owl monkey was well past the age when most neuroscientists believed the critical periods had closed. The existing dogma, as a result, held that the area of the somatosensory cortex devoted to processing signals from the severed nerve was likely to be lying dormant, unused, dead. Yet that was not at all what Merzenich found when they remapped the brain.

Instead when Merzenich and Kaas touched areas of the hand adjacent to those whose nerves had been cut off, the supposedly dormant parts of the somatosensory cortex lit up with activity. There was no mistaking it—the brain, supposedly long since hardwired and fixed in place, had somehow transformed itself once again. The areas of the hands that were active had colonized the abandoned neural areas.

How could this be? The results seemed to fly in the face of the work for which Hubel and Wiesel had won the Nobel Prize. And they were summarily dismissed by the field at large. But Merzenich continued to follow the thread. Next, he examined whether experience itself might prove enough to change the amount of brain space devoted to performing a specific task.

To do this, Merzenich once again mapped the area of the somatosensory cortex associated with individual fingers in a monkey. But this time, rather than cutting a nerve, a postdoc named William Jenkins spent three months training monkeys to master a task that would require them to develop an unusually fine set of skills. The team wanted to know if that too would be reflected in the brain. The task was an obscure one: Monkeys learned to maintain contact with a spinning disk, using the pads of two fingers, but to do so gently enough that the fingers could remain motionless yet not get dragged around with the disk. This required just the right amount of pressure.

Jenkins made sure the monkeys were highly motivated to learn the trick. Fail at the task and the monkeys had to wait to be fed until after training. Master the trick and the reward was untold riches, in the form of banana-flavored food pellets—as many as six hundred pellets over a twenty-four-hour period.

Jenkins's training rig, which he could affix to the front of the monkey cage, consisted of a 13-centimeter-diameter, pie-shaped aluminum disk, machined with a pattern of twenty alternating raised and lowered surfaces. A minute electric current ran through each monkey's home cage, flowed, unfelt, through their little fingers, and triggered a circuit every time they touched the metal disk. This in turn triggered the release of a pellet. Beginner monkeys usually collected their bounty from the pellet chute on the left side of their cage with a couple of fingers. The seasoned vets, however, learned to lick the pellets out of the dispenser, while leaving their fingers on the spinning disk, maximizing pellet flow.

As the monkeys got better at the task, Jenkins made them step up their game; he increased the duration the monkeys needed to keep their fingers on the disk to as long as fifteen seconds, speeding up the rate of its spin to as fast as one revolution a second, and finally moving the disk so far from the cage it was only possible to reach it with the tips of one or two of the longest fingers. The challenge at that point lay in applying just the right amount of pressure to the disk to maintain the electrical circuit, without slowing down the disk or allowing the fingers to be spun off by the force.

After more than a hundred days of training, Merzenich and Jenkins were ready to find out if this new kind of expertise had changed the layout of the cortex of the monkey's brains. The results were powerful. The amount of somatosensory cortex devoted to the sense of touch in the fingers used to skim the disk had increased 400 percent. The brains of the monkeys had rewired themselves based simply on practice motivated by the desire to maximize pellet flow.

Ever since Pascual-Leone had read Merzenich's monkey experiment while a young neurology resident, he'd wondered if he might find the same kind of phenomenon in humans. Eventually, he had an ingenious idea—why not study blind people who learned to read braille?

Like the monkeys with the spinning disks, braille readers used the pads of their individual fingers to perform a task requiring extremely refined sensory calibration. To read the columns of raised dots representing the letters of the braille alphabet, these individuals had developed what Pascual-Leone considered an "uncanny" ability to run their finger across the braille cells at a very rapid clip, while still discerning the presence of up to six dots per cell. Press down too hard with the pad of the finger, and the pace of reading would slow to a glacial clip. Press down too softly, and it would be impossible to tell how many dots were present. Instead, Braille readers moved their fingers multiple times over the dots in a blur of motion until they registered the code and moved on to the next.

"It wasn't the blindness itself, but the fact that blind people acquired braille reading that was of interest," Pascual-Leone says. "What kind of changes in the brain were going on?"

Pascual-Leone couldn't use single-neuron recordings. Human volunteers usually aren't willing to have their skulls sawed open and electrodes pushed into their brains. So instead he used a technique that delivered a weak electrical shock to the pad of each subject's reading finger. An array of electrodes affixed to the subject's scalp then detected where in the somatosensory cortex neurons fired.

When Pascual-Leone tallied his findings he discovered the preferred reading fingers of braille readers had indeed gained a far larger portion of cortical real estate than that of non-braille readers. What's more, just like in the monkey experiment, Pascual-Leone demonstrated that it came at the cost of cortical resources devoted to processing touch from adjacent fingers.

Next Pascual-Leone moved on to the motor cortex, demonstrating that the precision of movement required to read braille led to a similar process of reading-finger colonization in the area of the brain that controlled that finger's movement.

These were a groundbreaking pair of experiments—the first time this kind of plasticity had been shown in adult humans. But it was one of the follow-up experiments that really caused a stir and laid the groundwork for the line of research that would eventually lead to Pat Fletcher's trip to Boston.

Months after those initial experiments, one of Pascual-Leone's colleagues, Norihiro Sadato, wanted a broader picture of the way the overall motor cortex, the part of the brain that controls movement, changed with braille reading proficiency. So Sadato used a different brain-scanning technology, called "PET" (positron emission tomography), that took a more global picture of the brain. Sadato wasn't initially interested in areas of the brain outside the motor cortex. But the method delivered a snapshot of what was going on elsewhere nonetheless.

Pascual-Leone remembers vividly the day Sadato burst in with the results.

"Do you want the good news or the bad news first?" Sadato asked.

"Always the good news," Pascual-Leone responded.

The good news, Sadato said, was that he had gotten some good data from the motor cortex proving a hypothesis they had chosen to test in the study.

The bad news, Sadato told Pascual-Leone, is that "you are going to be completely uninterested in that."

When Sadato showed him the brain scan pictures, Pascual-Leone understood why. The motor cortex was indeed more active. But so was the visual cortex—somehow signals from the fingers were ending up all the way in the back of the brain, in an area it was previously thought could only receive such activation signals from the eyes. These blind braille readers were actually processing the words in the same area sighted people do when they read text using their eyes—in a sense, it seemed, the blind subjects were "seeing" with their fingers.

"I looked at the picture and said 'Oh, my God! What is that? Is that real? Is that an artifact?'" Pascual-Leone recalls. "And so that led to a lot of amazingly interesting sets of experiments and explorations."

Indeed, by the time Pat Fletcher arrived in Pascual-Leone's lab that day in 2006, he had followed the thread to some rather amazing discoveries. In 2000, Pascual-Leone learned of a sixty-three-year-old congenitally blind woman who had spent four to six hours a day for years reading braille as a proofreader in Spain. As a result, she was extremely proficient, reading faster than most sighted individuals can read a newspaper.

Then one day the woman complained of vertigo, passed out, and was rushed to the hospital. When she eventually awoke, the doctors informed her she was an extremely lucky lady. She had suffered two strokes. But miraculously, the doctors told her, the strokes had caused damage (killing neurons) only in the left and right side of her visual cortex—areas of the brain they assured her she didn't need, because she was blind. When the woman tried to read, however, she discovered that she could no longer decode braille. She could feel her way around it, perhaps identify a dot and use deductive reasoning to figure out the letter. But her fluency was gone.

To Pascual-Leone, who met with her, conducted brain scans, and published a case study, the experience seemed to confirm that the activity in the visual cortex during braille reading was most certainly not random. A few years later, Pascual-Leone had the opportunity to scan the brain of a congenitally blind Turkish painter named Eşref Armağan (http://esrefarmagan.com/). When he was a child, Armağan's family had owned a small store, and set their blind son outside every day so he wouldn't knock over the merchandise. To entertain himself, Armağan learned to draw patterns on the sand, feeling the pictures with his fingers. The drawings caught the delighted attention of passersby, eliciting praise and encouragement. With practice, Armağan's drawings became increasingly elaborate.

By the time Armağan reached adulthood he had developed a signature technique. He used a sharp pencil or braille stylus to etch his pictures in paper or a canvas, and trailed the other hand along the page behind it to keep track of where he was in the drawing. Remarkably, he could draw a detailed visual representation of an object after feeling it with his hands for just a few moments. He could also remember where the lines of his outline were, and fill in the spaces with colors—which allowed him to make oil paintings. Beautiful oil paintings, so lifelike and vivid they won him worldwide recognition. How was it that an individual who was completely blind could become such a talented visual artist?

In 2007, Pascual-Leone and the postdoc who ran his imaging lab,

Amir Amedi, had the opportunity to scan Armağan's brain when he came to the United States to present his work at New York City's Museum of Modern Art. In the laboratory, Amedi handed Armağan a series of objects and asked him to draw them. One of them was a figurine of a man sitting on a bench holding an apple. Armağan was able not only to draw it after touching it for just a few seconds—he was able to draw it from different perspectives, from straight ahead, from up above, from the side. The implications were profound.

"Touch is very different than sight," Amedi explains. "The concept of perspective is not relevant. Objects are the same size when they are closer and further away with touch. Yet without any visual experience, he had developed the ability to mentally create a three-D representation of an object in his brain and manipulate it, so he could draw it from any angle. He did so with a precision and quickness that would be difficult for even a sighted person to do. I was amazed by him."

Next Amedi and Pascual-Leone scanned Armağan's brain while he lay in an MRI machine and drew on paper placed on his stomach. As he drew, considering the object from different perspectives, and bringing it to life on the page, Armağan seemed to be seeing and manipulating the object in his mind's eye. He too was relying on the visual areas of his brain, which were alight with activity.

"In order to do that he uses the same circuitry devoted to seeing by the sighted, a slightly different order, but the same circuitry," Pascual-Leone says. "Is that seeing? He doesn't 'see.' But if you were to be looking at brain activity you would say it is not fundamentally different."

By the time Pat and Adam came into the lab for their tests, Pascual-Leone and Amedi had spent a considerable amount of time discussing what they hoped to explore. One of things about Pat Fletcher that Pascual-Leone found particularly wild was her claim that she could both "see" and "hear" at the same time, using her ear to gather the sensory information she used to experience both sensations. And indeed, when Pat entered the laboratory that day, she readily demonstrated her ability to spot the door or look around, while simultaneously carrying on a conversation.

"That is absolutely amazing to me," Pascual-Leone says. "One implication would be that there might be truly, truly separate neural substrates processing these streams of information, even though both get in through the ears. So we set out to test that."

Sure enough, when the researchers played Pat regular sounds, like a whistle, the normal areas of the brain associated with auditory processing lit up. When they played her soundscapes, her visual system also activated. When the researchers scrambled up the soundscapes so as to make them gibberish, the occipital lobe remained inactive and Pat reported she could not see anything.

Somehow, Pat's brain was able to discern the soundscapes from other sounds—and route them to the area of the brain associated with identifying objects.

It was a memorable day for both Pascual-Leone and Amedi, who years later still marvel at the abilities Pat showed off. But when Pat herself thinks back on it, one of her most vivid memories is not in the lab itself, but in a shopping mall afterward.

Back at the lab, Amedi hadn't quite been ready to confirm that what Adam and Pat were experiencing was in fact "sight"—to do so made him uncomfortable. "As a scientist I have to be a bit careful," he says now. "I can't say 'sight.' But there is no question that they are actually recruiting the same system. So we told [Adam] the visual system was activated."

It was enough for Adam Shaible, whom Pat vividly recalls watching with her vOICe system as he "practically danced" gleefully past the shops in the shopping mall, ecstatic that scientists at Harvard had finally verified his own experience.

"He was so happy to have it confirmed that 'yes, he does have sight,'" Pat says. "It was just so cool to watch him, and to hear his happiness and to hear his confirmation, and to understand how much it meant to a person who had been blind all his life that they actually had confirmation they actually could see. To me that is one of the greater experiences."

So how to explain the discrepancy between Hubel and Wiesel's critical periods and Pat's experience? How to reconcile the findings of that kitten permanently blinded in one eye with the more hopeful work of Mike Merzenich and his patients who learned to hear again using cochlear implants?

Over the last decade, biochemists have begun to unearth some answers that help reconcile all these contradictions, and offer a more nuanced view of the rules governing critical periods and neuroplasticity.

It's impossible, after all, to deny that we are all capable of lifelong learning. At the same time, who would deny that the brain of a child is more supple and open to learning new things than that of an adult? Certainly not anyone who has ever tried to learn a second language in adulthood and failed to shed those last traces of a telltale American accent. There's a reason we often refer to five-year-olds as "sponges" and marvel at their ability to soak up information.

Indeed, almost from the moment Hubel and Wiesel first demonstrated that critical learning periods existed, scientists began searching for ways to hack the system and restore the plasticity to the adult brain that is present in a child. Some suggested that if we could explain why critical periods opened and closed, we could enhance learning—even invent "learning pills."

From the start everyone assumed the key was to add something to the brain, such as the stem cells or growth factors we learned about in the last chapter. Just like those muscles and cartilage cells

Stephen Badylak and Gordana Vunjak-Novakovic regrew, perhaps regeneration was the key. But in recent years, scientists have begun to realize that the key to reopening critical periods is, in fact, not to add something at all—the key, surprisingly, appears to be to take something away.

As we have learned, neuroscientists have long accepted as dogma the idea that neurons that *fire together, wire together.* But there are many factors that can make a neuron more or less likely to fire, and, apparently, to wire to its neighbors, according to Takao Hensch, a Harvard neurobiologist.

As we age, Hensch and others are discovering, biochemical processes occur that create molecular "brakes" on plasticity, dramatically inhibiting the ability of neurons to form new connections with their neighbors. These molecular brakes don't prevent new connections altogether. But they dampen the impact of chemicals that, when released in a child's brain or the brain of those albino tadpoles we learned about, either cause neurons to fire more easily or cause those neurons to more promiscuously form new connections to one another.

Behaviorally, one's enthusiasm for seeing a new kind of truck or a princess castle fades as we age simply because these things no longer seem as novel to us as they might to a three-year-old. But it's also true that the loss of youthful enthusiasm is reflected in very real structural changes in the brain.

"The systems in kids are naturally revved up about just about everything because they are interested in learning how the world works," Hensch says. "But as we get older we kind of get jaded perhaps. In biochemical terms, our systems are less easily engaged."

That doesn't mean, however, that plasticity fades altogether, Hensch says. When we are deeply immersed in something—one of those "brain training" video games, for instance—the areas of our brain that regulate attention and focus can flood other parts of the brain with chemicals, called "neuromodulators," that make the neurons in those areas more likely to fire. The modulators put those neurons on alert, in other words, and prime them to respond to other neurons firing in their environment. Pat Fletcher's boundless enthusiasm and deep focus, as well as those hours spent practicing with her vOICe system, no doubt mobilized many of the neuromodulators available to her. Over time, new connections formed in her brain. It was a triumph of her curiosity, focus, and will.

But it turns out that as we age, the body also begins to produce substances—and sometimes build physical structures—that dampen the impact of these modulators. They can lull certain populations of neurons into a state of lethargy, or simple disinterest. Stroke victims can learn to rehabilitate. Pat Fletcher can learn to see with her ears. But it is an uphill battle—a battle against the mature body's built-in predisposition to defend the infrastructure as it exists, infrastructure that was built slowly, deliberately, and at great cost over years.

One of the watersheds that led scientists to suspect as much occurred in the early 2000s, when the Italian biologist Lamberto Maffei decided to borrow a page from the field of regenerative medicine and apply it to neuroscience.

For centuries scientists have wondered why it was that we can regenerate peripheral nerves in the body, but are incapable of regrowing axons capable of conveying electrical impulses from the brain, through the spine to the outer limbs. This mystery has condemned thousands of individuals with spinal cord injuries, such as the late actor Christopher Reeve, to life in a wheelchair.

In the 1990s and early 2000s, some leaders in the field of regenerative medicine began to close in on at least one answer. The body, it turns out, creates proteins called "CSPGs" (chondroitin sulfate proteoglycans), which develop as the body matures and, when present, impair the growth of adult axons. In a healthy adult these molecules serve an important purpose—signaling that the body is mature and should stop changing; that the appropriate structure is in place, and should now be protected as is.

They also play an important role in protecting the body when it is injured.

Once these axons are severed, however, as was the case for Christopher Reeve when he was brutally thrown from a horse in 1995, the presence of CSPGs becomes a tragic disadvantage. If scientists could find a way to destroy these compounds, might the axons begin to grow again? In fact, by designing enzymes that could destroy CSPGs, scientists experimenting on paralyzed rats were able to show just that.

In his lab, Maffei wondered if the same mechanism might be in play in the brain. After all, the brain cells involved in vision, hearing, and, in fact, all cognitive functions also rely on axons. Maffei conducted the same experiment performed by Hubel and Wiesel with the kitten: He sewed one eyelid of a rat shut and allowed the second eye to develop normally. Just like that of Hubel and Wiesel's kittens, Maffei's rat's vision remained significantly impaired even after he had removed the sutures.

But then Maffei modified ideas from his colleagues in the field of regenerative medicine. He injected the bacterial enzyme that destroyed the CSPG molecules directly into the rat's visual cortex and something amazing happened: The rat began to recover vision in the blind eye. Maffei had reopened a critical period. He had removed the brakes and increased plasticity.

In the visual cortex, Hensch explains, CSPGs form what he calls "perineuronal nets" (PNN). They wrap like a "glove" or "Saran wrap" around the neurons, preventing those branching projections from bumping into the dendrites of other neurons and forming new connections. By destroying these coatings, Maffei freed the brain cells to bond again.

Other groups have since identified more kinds of molecular brakes. Over time, myelin sheaths, the lipid coatings that surround axons, can become covered in proteins that coat them like barnacles on a ship, preventing new projections from sprouting or making contact. When researchers at Yale created mice with mutations that prevented them from developing one type of protein called a "Nogo receptor," the mice matured with open-ended critical periods.

In his lab, Hensch has begun to examine the possibility that the genes that code for all of these proteins can be manipulated to turn plasticity up or down. He has, in other words, been moving ever closer to demonstrating how one might create that "learning pill." Remarkably, in 2013 Hensch demonstrated that by feeding volunteers Depakote, a drug commonly used to treat mood disorders and epilepsy, and then assigning them to perform drills on a computer, he could in just two weeks dramatically increase their ability to learn a skill that usually can only be acquired in childhood—the ability to identify a specific musical note without first being offered an example of another note to compare it to. This skill is known as "absolute pitch."

"To my knowledge this is the first case where we have been able to change or acquire perfect pitch in adulthood," he says, "certainly in a two-week period. These manipulations offer a possibility of change. But you still have to work at it to make the change happen.

"The biology has reached a point where we shouldn't give up and accept that plasticity is gone forever," he says; it's "simply that it can be highly regulated in adulthood. Closing critical periods seems to be just as important as having them opened. And so finding ways in which to lift the brakes is going to become very important."

Even without the chemical interventions, it remains possible for the brain to change. These "brakes" can be overcome, we have seen, with sufficient repetition and practice—just as it's possible to knock down a piece of wallboard if you hit it with a hammer enough times.

But Merzenich's cochlear implant patients still had to wield that hammer many times, and to wait for months for their brains to forge new neural pathways and learn to decipher the patterns of electrical signals representing sounds. Pat Fletcher was able to adjust to her soundscape machine relatively quickly because some of the old visual pathways still existed from the twenty-plus years she had lived as a sighted individual. But it still took months more for her brain to learn to interpret the signals in three dimensions.

The case of Pat Fletcher "is a wonderful demonstration that we should never give up simply because of age," Hensch says. "The fact that [Pat] had been deprived of input to that part of the brain which normally processes vision but then [was] allowed the opportunity to innervate it with something else suggests that even in the adult case there is the opportunity for remarkable plasticity if you find the right neural conditions."

Hensch suggests there may someday soon be a way to artificially accelerate that process, and that it could be used to speed up or reopen any number of different learning processes—the ability to learn a second language without an accent, for instance. The ability to soak up information like a sponge. The ability for someone like Pat Fletcher to rewire her brain and see with her ears. Rehabilitation for adults who have had strokes.

In the years since the initial experiments with Pat Fletcher, Pascual-Leone's protégé, Amir Amedi, has opened his own lab at Hebrew University in Jerusalem, and he has expanded the findings to dozens of blind individuals. He has also expanded his research and discovered that the same area of the brain that processes objects also appears to be associated with color. Amedi also has unveiled a new sensory substitution device he calls "eye music," which allows blind people to "see" these colors by assigning a different instrument to each one. Over time, the brain learns to associate the different timbres with different colors.

Now Amedi is considering the next logical step: how we might be able to use this area of the brain for augmentation. Amedi likes to invoke an image of James Bond wearing a vOICe camera mount and headphones on his ears. Instead of a camera, however, Bond's vOICe system is hooked up to an infrared or thermal sensor.

"I like to call the idea 'For Your Ears Only,'" Amedi says. "We can start using the sensory substitution to augment rather than to substitute for the missing sense."

Imagine Bond entering a building and heading down a hallway, toward the lair of an evil mastermind. Using his regular sight, Bond could scan the hallway in front of him for guards and attackers in his line of sight. Meanwhile, 007 could use his vOICe system to see through the walls and detect the heat signatures of lurking bad guys waiting to jump out and surprise him. He could even shoot them through the walls before they could see him.

Amedi has only just begun to conduct research on what areas of the brain would be recruited to process this information. But the possibilities are intriguing. And they are really just a beginning. The plasticity of the brain, it turns out, doesn't just help the body heal itself after an injury. It also seems to play a role, believe it or not, in forming our ability to protect ourselves in dangerous situations. Wisdom somehow stored beneath the consciousness of our brain in the strength of synaptic connections molded by experience might, in other words, help us to avoid injury all together. Thus plasticity, in addition to explaining the obscure mystery of Pat's vOICe machine, might also explain something that all of us have likely experienced but that can seem so fleeting, so unexplainable that we often dismiss it as entirely of our own imagination. Plasticity, we shall see, also might explain one of the great mysteries of human experience: intuition.

CHAPTER 5

SOLDIERS WITH SPIDEY SENSE

INTUITION AND IMPLICIT LEARNING

The young fire lieutenant led his hose crew into the one-story, single-family home in Cleveland and headed toward the blaze in the rear kitchen area. They attacked with water, but the flames just roared back. After several attempts they retreated to the living room to regroup. It was uncanny, the lieutenant was already thinking; the water should have had more of an impact.

Then, as they stood there deciding what to do, the lieutenant got a feeling—the kind that sends a cold chill down your spine or makes your hair stand on end. He couldn't put his finger on it. He couldn't say what was wrong. But he felt it with the same powerful certainty you or I might feel when we spot a menacing face sneering at us from across a crowded subway platform or see an errant car speeding straight toward us—he and his men *had* to get out of there.

Just as they left the building, the floor where they had been standing collapsed. Flames had been raging unseen and out of earshot beneath them in the basement, their danger camouflaged by a thick living room floor that had muffled the roar of their destruction.

If the lieutenant hadn't ordered his men out, they all would have plunged into the fire below. After that day, the lieutenant became a firm believer in extrasensory perception (ESP). "A sixth sense," he would later insist to Gary Klein, the research psychologist who discovered his case, was part of the makeup of every skilled commander.

"According to everything that's known, this should be impossible," Klein, an affable researcher with a mischievous smile, told me of the many cases of life-and-death decisions he detailed in his 1999 book, *Sources of Power.* "It's unconscious, it's intuitive. But it isn't magical. People can do it—they make these kinds of decisions under time pressure and uncertainty all the time."

Once Klein began looking, he found examples of this kind of "sixth sense" at work in life-and-death situations over and over again, for instance in the actions of Lieutenant Commander Michael Riley, an air defense officer on the British destroyer HMS *Gloucester.* At 5 A.M. one day near the end of the first Persian Gulf War, Riley spotted a blip on a radar screen headed straight toward his ship. Experts would later insist the object was virtually the same size and was moving at the same speed as the U.S. Navy warplanes that had been flying over Riley's ship for days. There was, they claimed, simply no way to know it wasn't a friendly. But Riley had a feeling—more than a feeling. Within seconds of its appearance on his radar screen, somehow he *knew* that that blip was a Silkworm missile, a projectile the size of a school bus, headed toward him.

"I believed I had one minute left to live," he would later say, though for the life of him, Riley himself could not explain how he knew. Riley did something he normally wouldn't do; he checked a second radar that measured the object's altitude. And when it showed that it was flying lower than an airplane normally would fly, he shot the object down. After four tense hours, it was finally confirmed—Riley had been right. He had saved many lives. Just as Pat Fletcher "saw" without using her actual eyes, Riley had somehow seen this

Silkworm missile, or at least intuited something concretely menacing in his mind's eye—without actually consciously seeing it.

But how was that possible?

We've probably all experienced something like this at one time or another. A "Spidey sense." A sudden sharp intuition—one of those unexplainable feelings that makes one snap to attention, or produces goose bumps. Intuition can seem otherworldly, divine even. But it can be a maddening, unnerving experience, because often when these feelings hit us, we have no idea why.

What's behind that feeling? How is it that that fire commander in Cleveland and Michael Riley just *knew* their lives were in danger?

It's one of the questions that Klein, a pioneer in a field of study known as "naturalistic decision making," has spent decades attempting to answer. And the answer, he has discovered, has nothing to do with ESP. When pressed, the fire commander recalled that the floor he'd been standing on had been too hot, and the flames in the kitchen were too quiet to account for the intensity of the heat. Even though he wasn't consciously aware of it, even though there wasn't time for him to realize it, the lieutenant's senses detected that things didn't add up. It was this discrepancy—what Klein calls a "pattern mismatch"—that made the firefighter's hair stand on end.

Riley, Klein would later suggest, had picked up on an almost imperceptible difference in the way the Silkworm missile radar blip had first appeared on his screen. That difference had been caused by the missile's slightly lower altitude. He hadn't been consciously aware of it. The discrepancy had lasted less than a second; it was too fleeting to break through to his conscious mind. But Riley's well-trained senses knew something was wrong. So he got that feeling.

Klein's findings provide a powerful illustration of an age-old truth with wide implications for all of us, a truth any hardened combat vet has likely experienced firsthand: that our conscious awareness at any given moment often captures just a small fraction of what's really going on in our environment—but other parts of our brains are furiously ticking through this flood of sensory information and processing it. Much of that information never makes it into our conscious awareness.

We know from Mike Merzenich and Pat Fletcher that there are many ways to get sensory information from the outside into our brains—that it's even possible to upgrade those sensory collection tools. But what about what happens in the brain once that information arrives?

Imagine if we could somehow tap into the areas of our brain ticking through sensory information outside our awareness, and train ourselves to better use it? We could improve performance and increase intelligence. We could even save lives.

The roar of trains leaving the station echoes off the walls of the concrete tunnel around me as I step onto an impossibly long escalator that hefts me slowly up from the depths of the Washington, D.C., Metro and lets me off outside on a crisp, autumn morning. As my eyes adjust to the blinding Northern Virginia sunlight, they are drawn to a hulking figure with toned biceps in military fatigues standing a few yards ahead of me. He is holding a submachine gun and is watching me.

I've arrived at the Pentagon, the seat of U.S. military might. But I'm not here to meet any of the usual suspects, one of those square-jawed generals you see on TV or some no-nonsense military strategist. I've come to meet a rarer breed of soldier, members of a small, elite band of pointy-headed Pentagon geeks who most people don't even know exist. They are the Pentagon's research psychologists and neuroscientists, and in recent years they have embarked on an audacious and tantalizing goal—they are searching for ways to help U.S. soldiers hone their "Spidey sense."

The surprising, superhero-like quality of an engineered "sixth sense" is part of what has led me here. But I've been lured by something more than pure intellectual fascination. So far, I've met a number of amazing individuals, harnessing science to adjust to their circumstances and bounce back from setbacks I can only imagine—industrial accidents, severe frostbite, artillery attacks, car crashes. It's inspiring. But the truth is, it's also a little alarming at times. These incidents remind me continuously of my own mortality, of the fragility of the human experience—which makes the idea that it might be possible to find ways to avoid these calamities altogether all the more seductive.

Hugh Herr's bionics, Lee Sweeney's muscle tweaks, and Pat Fletcher's soundscapes all hint at untapped powers already inside us. It makes sense that we might also have powers not just to heal ourselves, but to protect ourselves. If there's something accessible to all of us already there in the back of our minds—a wisdom, a life-force gleaned from the sum of our experiences—I wonder if there's a way for it to guide us not just through or away from dangerous situations, but anytime we are paralyzed by indecision or feel lost.

It's no coincidence, of course, that the obscure geek squad behind some of the most promising efforts to solve the mystery of intuition consists of soldiers. Battle is a dangerous and deadly pursuit. As if to drive home the point, I watch as the man entering in front of me at the main entrance's metal detector offers to take off a prosthetic arm before stepping through.

The sixth sense of the seasoned warrior is something that has been prized since the first two cave clans met on the open planes and attacked one another with clubs back in the Stone Age. During the Vietnam War, stories of Rambo-like special forces operatives able to sniff out danger and halt patrols within just a few feet of waiting ambushers circulated so often they were almost a cliché.

Iraq with its IEDs hidden in plain sight and the craggy ambush-friendly terrain of Afghanistan have also proven ideal testing grounds for the existence of intuition. And when I begin to investigate, I learn that sure enough, almost from the beginning of the wars there, soldiers traded stories in hushed tones using words like *intuition, Spidey sense,* and *ESP* to describe near misses, last-minute saves, or that one guy in the unit who always seemed to know when the shit was about to go down. It wasn't long before the Pentagon's scientists began to take notice.

"You hear it often," says Peter Squire, an experimental psychologist working with the navy who is leading the current project on intuition. "There's usually one person in a squad, or one guy people remember who just had this ability or hunch to know where IEDs were. These individuals could just go out there and sense by surveying the environment that something was amiss and that they should stop. And in fact when they went out then and did survey or reconnaissance on the area, they would identify an IED or something."

One afternoon in 2006, then–Lieutenant Commander Joseph Cohn, a Brandeis-trained neuroscientist and Squire's predecessor, was at the Naval Air Warfare Center in Orlando, Florida, chatting with a marine colonel about the subject. There'd been several high-profile accounts in the media in recent months—a sergeant in Iraq, for instance, who had anticipated a bombing outside the base's Internet café while chatting with his wife via cell phone. The sergeant had noticed a man outside and got one of those "feelings." So he'd watched closely—and pounced when that man planted a package and ran away, and shouted for the patrons to evacuate the café. Or the story about the survivors of a Canadian platoon ambushed in a marijuana field outside a schoolhouse in Kandahar, Afghanistan. Some of them swore their "Spidey senses" had been tingling just before the Taliban rockets rent the early morning quiet.

That day in Orlando, the colonel said he had little doubt that these accounts were worth listening to. In fact, he told Cohn, there was a sergeant in *his* unit with a sixth sense so highly tuned that everyone always wanted him out on patrol with them.

"He always seemed to know when to start ducking, when to

start shooting, even before things would start," the colonel said to Cohn. "If we were going to clear buildings, he was the one that would get the sense that things were going to turn south. And he would be the one who would tell folks to take cover."

Then the colonel said something that got Cohn thinking. "Can you do something like that, Doc?" he asked Cohn. "Can you make people be able to do that?"

It seemed a far-fetched idea. But, as luck would have it, Cohn was in the business of looking for far-fetched ideas.

Cohn is sitting in a cramped conference room deep inside the labyrinthine bowels of the Pentagon when I meet him and he tells me this story. A stocky sailor with a salt-and-pepper buzz cut and an unlined face, Cohn is wearing his navy tans. He's just completed a mandatory physical fitness test. He jokes that he would have failed the weigh-in if he'd eaten breakfast that morning. But Cohn is trim and fit, like most of the soldiers and airmen I passed in the winding, fluorescent-lit hallways on the way to meet him.

Cohn graduated from Brandeis University with a Ph.D. in neuroscience in 1998—an expertise he has been using ever since to understand and improve the minds of America's servicemen and women. Back in the mid-2000s, he was a program officer at the Defense Advanced Research Projects Agency. DARPA's stated mission is to fund research for national security ideas that many think to be impossible—the harder, in fact, the better.

The job of a DARPA project manager is to pose provocative questions in the hopes of developing new technologies not five years out, not ten years out but far beyond immediate military needs—so far beyond, so challenging, that few other agencies or companies in the private sector would be willing to fund it.

DARPA laid the groundwork for the Internet and developed the Global Positioning System, or GPS. DARPA is funding limb regeneration projects and some of the most far-out-there work with prosthetics, neuroplasticity, and brain–computer devices.

Might it be possible, Cohn wondered, to actually identify and "quantify" intuition? If it is something that is really happening in the brain, it seemed to him, then surely there must be a way to track it—to see that it is missing in some people, or present in others. To watch it in real time. It was an exciting idea, because "if you could do that," Cohn says, "then you could figure out ways to train it."

"How do you take someone who grew up in Montana and put him in some foreign country in the Middle East and make it so that he is suddenly able to pick up the cues that are telling him something is going wrong?" Cohn says. "How do you do that, even though the person has never seen the place and hasn't been raised to understand those cues or make predictions based on those cues?"

Over the course of his career, Cohn had actually overseen the funding of a number of studies on intuition that had produced interesting results—results that suggested intuition was indeed something tangible. After the conversation with that colonel, Cohn went back to the literature and refreshed his memory. He also thought back to some of the researchers he'd previously worked with. One in particular came to mind—a guy who'd done groundbreaking work with firefighters in the field. That researcher's name was Gary Klein.

Yellow Springs, Ohio, is a long way from the Pentagon. It's a tiny speck of buildings—its downtown covers a total of just a few blocks—surrounded by a small outer layer of picturesque farmhouses, lost in a vast sea of cornfields. Miles and miles of cornfields, a vastness suggestive of possibility and mystery, crop circles, and strange doings.

The little town itself seems far bigger than its diminutive, geographical footprint. Since the 1850s, it's been perhaps best known as the home of one of the nation's most liberal colleges, Antioch, haven for long hair and social activism. And it remains a favorite weekend destination for city folks looking for the midwestern version of Woodstock. The comedian David Chappelle has a compound outside of town.

For many years, it was also something of a hot spot for intuition research, thanks to the presence of its longtime resident Gary Klein.

Klein didn't set out to study intuition. The bearded psychologist was initially interested in what seemed an entirely different question. He wanted to know how people made really hard decisions under extreme time pressure and uncertainty. How did they manage to function, in other words, under extreme stress?

It was the mid-1980s. Back then, everything Klein had read suggested that individuals shouldn't be able to make rational decisions in these types of high-pressure situations. These kinds of scenarios often call for snap decisions made with limited time to weigh appropriate options. High stakes and stress hormones have the power to overwhelm even the most relaxed individual. Yet police officers, stock traders, and experienced military commanders seemed to make decisions—the right decisions—under pressure all the time. What was it that set them apart from those who froze like a wild animal caught in the headlights of an oncoming car?

It was, Klein knew even then, a question of extreme interest to the U.S. military. After earning a Ph.D. in experimental psychology from the University of Pittsburgh and entering academia, Klein had gone to work as a research psychologist for the U.S. Air Force. He founded his own research firm in 1978. Doing work for the military, where he still had excellent contacts, seemed the obvious path. And in the mid-1980s, many of his contacts were asking questions about decision making in military terms. What actually happens in a battle, they wanted to know, when a mission falls apart and you've got to completely improvise? What if you encounter a surprise—like an enemy base camp in the middle of the jungle that you didn't know about, housing a unit far larger than your own? What goes into making the right decisions on the fly—and how can you train individuals so that they make better ones?

The Army Research Institute for the Behavioral and Social Sciences, in particular, had funded a number of studies aimed at answering precisely these questions. Klein's contacts all told him the same thing: The findings were disappointing, the recommendations useless. The problem was that the experimental conditions favored by academics didn't have much to do with the conditions real soldiers might encounter out in the field. What did a bunch of college sophomores sitting in a tightly controlled laboratory answering questions have to do with a soldier who stumbles into an ambush? Or a commander under extreme pressure who suddenly gets a new vital piece of information—and instantly changes his plan?

So when the army finally published a new call for researchers to examine the problem again, Klein wrote a proposal advocating an entirely different approach. Instead of a strictly controlled laboratory setting, Klein suggested something far messier: heading out into the field and watching professionals who seemed to defy the norm. Only then would he try to figure out what it was about their approach that set them apart.

"I didn't want to do the same kind of research everyone else was doing, because they were looking in the wrong place," he says. "There's a real mastery here. According to everything that is known this shouldn't be possible—these people shouldn't be able to make these life-and-death decisions under time pressure and uncertainty. And yet people can do it."

Klein suggested the army finance a study of firefighters. Soon Klein and his newly funded team hit the road visiting fire stations across the Midwest, in places like Dayton and Indianapolis, riding along to big blazes and sitting down with seasoned commanders in back offices to listen to their war stories. Cleveland was a favorite destination because it had a sizable chunk of decaying housing stock, more likely to catch fire.

When he started out, Klein had a pretty good idea about what he might find. He suspected that expert commanders picked a limited range of options to choose from and then carefully weighed the pros and cons. Under extreme time pressure, they might choose only two or three, with a favorite option they could compare to the others before taking action. But the initial approach was still likely in play. In other words, Klein, like most people, like the psychological literature that existed at the time, expected a logical approach to unfold in every commander's conscious mind.

Suggesting a commander might consider only two options, Klein thought at the time, was a rather "daring hypothesis," since it deviated from the accepted idea that we cycle through many different options before making a decision. But Klein soon discovered he was in fact too conservative. Time and time again Klein found the same thing: The commanders only looked at *one* option. They just "knew" what to do. They didn't consciously weigh the pros and cons at all—at most they simply imagined the scenario unfolding. By the time they became aware of the approach, they'd already decided it was the best one. They operated entirely on imagination and instinct.

"It really shook us because we didn't expect that," Klein recalls. "How can you just look at one option? The answer was that they had twenty years of experience. That's what twenty years of experience buys you."

Twenty years' experience, it turned out, bought the firefighters the ability to do what Klein called "pattern matching." It was a process that seemed to take into account a range of sensory information that somehow occurs without any conscious deliberation. Instead of weighing the options, the veterans simply said, "I know what's going on here, and because of that, I know what I should do." After that first approach popped into their head, the commanders didn't compare it to others. They evaluated whether it would work by imagining it unfold. If they found a flaw they moved on to the next best option, which also popped seemingly effortlessly to mind.

"It was unconscious, it was intuitive, but it wasn't magical," Klein says. "You look at a situation and you say, 'I know what's going on here, I've seen it before, I can recognize it.' All of the conventional researchers had been looking at college sophomores, giving them tasks they had never seen before. So there wasn't any experience. So they missed it."

There's a corollary to Klein's model that helps explain how that firefighter at the beginning of the chapter knew to get his men out of the burning house, how Lieutenant Commander Michael Riley of the HMS *Gloucester* shot down that Silkworm missile, and how that sergeant in Iraq always knew when to duck.

The brain's unconscious pattern-matching machinery can help connect a solution to a problem. But Klein argues that it can also detect mismatches—anomalies that alert us when something out of the ordinary is taking place. That something is wrong. That's why we get a "Spidey sense," and that's how that fire commander in Cleveland knew he had to get his men out of that burning house moments before the floor collapsed. The commander didn't consciously know that the flames in the kitchen weren't big enough or loud enough to account for the heat around them—he couldn't have told you that at the time. He just had a "feeling." He just knew "something was wrong"—so wrong it made his hair stand on end. Klein believed that somewhere else in that firefighter's brain the sensory input was being processed and producing the "pattern mismatch" that was setting off alarm bells. Back then, Klein didn't have the brain-scanning technologies we have now to watch this unfold in real time. But his hypothesis was powerful nonetheless.

Klein's insight was that expertise gave rise to intuition—that, in fact, intuition (when it is right) is really just another form of unconscious knowledge. This was the way Klein thought about it, the words he used to describe the problem, because he was a trained psychologist, not a neuroscientist. But in a sense Klein had stumbled on the very same insight at the heart of the work of Mike Merzenich and Alvaro Pascual-Leone.

As we have seen, Merzenich concluded that the brain is essentially an elaborate pattern-recognition machine. It is dynamic and ever changing, and capable of amazing feats of associative learning. In the case of Merzenich's cochlear implant, the brain learned to associate specific pulses of electrical stimuli with specific sounds and words meant to represent ideas in the outside world. It adjusted to a new "front end" on the auditory system. These adjustments were reflected in the very physical structure of the brain. In the case of the firefighters, Klein seemed to be suggesting, the brain learned to associate specific sensory stimuli, such as heat and sound, with different external conditions, such as the danger posed by a raging cellar fire masked by a thick living room floor.

It was the pattern-matching abilities of Klein's firefighters experts that allowed them to make decisions on the fly, and to detect anomalies that set their Spidey senses tingling. It was the pattern-matching abilities of Merzenich's deaf patients that allowed them to hear using consistent patterns of electrical pulses. Remarkably, in both of these examples—the translation of electrical stimuli into sounds with meaning, and the translation of sensory stimuli into an intuition of danger—these associations were formed and resided somewhere *outside* the conscious mind.

When Joseph Cohn read Klein's research on firefighters it immediately raised an obvious question. If a "pattern mismatch" was indeed setting off the alarm bells in the brains of Klein's firefighters and that sergeant the colonel had told him about, what was actually happening in the brain? Was it indeed possible that such a high-level analysis could occur *outside* the conscious areas of the brain associated with intelligence and human consciousness? What did that pattern mismatch actually look like in the brain? Could you detect it?

Cohn knew that if he could answer these questions, it would

open a number of exciting possibilities. Once he found a neural signature of intuition, the military might actually be able to train it and harness it to help enhance decision making.

But how, he wondered, would you even begin looking?

In the movie *Limitless,* Bradley Cooper plays a disheveled thirty-five-year-old writer who has just been dumped by his girlfriend, is paralyzed by self-doubt, smokes too much, and seems to have missed "the on-ramp" to life. He is, in other words, working well below his potential. Then he takes a magical pill called "NZT," which allows him to access the "eighty percent" of his brain that is usually off-limits. He finishes writing a book in four days. In less than a week Cooper has transformed into a slick, well-groomed, genius stock picker. By the end of the movie he is about to win election to the U.S. Senate, and an eventual presidency seems assured. Personally, when I saw the movie, I thought this was a guy I could root for and relate to. And I found it hard not to wonder, was there perhaps a potential presidency in me?

Alas, Commander Cohn informs me, the idea that we normally only use 10 percent (or in the case of *Limitless* 20 percent) of the brain is a fallacy. The truth, Cohn's very first neuroscience instructor at Brandeis insisted, is that there has never been a way of quantifying total brain activity in any reasonable way. Even if there was, the idea just doesn't make evolutionary sense. Nature is ruthlessly efficient. If we didn't need 100 percent of our brain cells, they wouldn't have survived the crucible of natural selection. Brain cells require too much energy to sustain to sit permanently inactive.

But there's a reason that the 10 percent myth has gained a foot-hold in the popular imagination. (Another recent movie, *Lucy,* starring Scarlett Johansson, also draws on it.) Much of what goes on in our brain *does* in fact occur outside our conscious awareness—a truth

I learned in my first-year psychology class during a basic lecture on the work of Sigmund Freud. Freud focused on repressed traumatic memories or emotions outside our awareness that cause us to, for instance, develop an unexplained phobia, or a strange obsession.

Cohn was interested in a very different kind of neural activity outside awareness, unconscious perception, which also has a rich literature. The human visual system, in particular, scientists have known for decades, is capable of registering information at a dizzyingly rapid rate that far exceeds our capacity to consciously process it all. How fast? Believe it or not, theoretically the brain can register images at a pace of 36,000 an hour. Or put another way, that's a pace of about 864,000 images over a twenty-four-hour period.

Harvard scientists clocked that rate in the 1960s by stringing together pictures cut out of magazines into a short film, instructing test subjects what to look for, and then beaming the film onto a screen using a 16 mm projector running at a high-speed setting. The scientist who invented the technique, a pioneering visual researcher named Mary "Molly" Potter, called it "rapid serial visual presentation" (RSVP). If Potter told her subjects beforehand what images to look for—a scene at a restaurant, a red fire hydrant, or a fish in a fish tank—she discovered, they could discern the presence of the image within the blindingly fast visual stream, even when the projector was running the film at its top speed of 10 images per second.

But if the subjects weren't told to look for a specific picture beforehand, the subjects often didn't *consciously* register the presence of many of the pictures at all. It was not, however, that they didn't see them. The first part of Potter's experiment had demonstrated they were perfectly capable of seeing pictures and reporting their presence when told beforehand to look for them. But if they weren't actively looking for an image, the brain quickly let that image go. The passing images registered for a moment—if that—on an ephemeral form of memory that she calls "short-term conceptual memory." Then the brain let them go and they were gone.

"Right away we knew that, 'Gosh, yes, they do understand pictures when they're coming at that rate, but that understanding is fleeting,'" Potter explains of her results. "They just cannot hang on to it. If you only had that one more second on that picture you'd remember it next year. But you don't, so it's gone in an instant."*

With that knowledge in mind, it's easy to explain how Michael Riley was able to spot that missile. It's also easy to understand why the Pentagon would be interested in somehow exploiting this ability. And, in fact, Cohn is by no means the first Pentagon scientist to attempt to do so.

"The brain knows more than you think you do," Amy Kruse, a former Department of Defense neuroscientist who played a key role in developing a number of projects that preceded Cohn, told me. "Our sensing system evolutionarily speaking has really been designed to be fast, right?† It's our cognition and all of the baggage that's associated with that 'thinking' part of our brain that essentially slows us down."

In the early 2000s, Kruse was among the neuroscientists working on another DARPA program called "augmented cognition," or AugCog, which resulted in more than $100 million worth of research into different ways the DOD might make warfighters "smarter"—an effort that created technology Cohn would later use in his project.

AugCog was originally headed by a navy behavioral scientist,

---

* Remarkably, Potter, now in her eighties, recently published new research using even faster modern technologies that demonstrated some visual processing occurs after we see a picture for just 13 milliseconds—or less than one seventy-fifth of a second. (That's a rate of 270,000 images an hour, or 6.5 million images in a day.) Though the ability to detect pictures at that rate was above chance, however, it was by no means 100 percent. Often times, in other words, we miss some pictures at that rate.

† For more on this, see Daniel Kahneman, *Thinking, Fast and Slow* (New York: Farrar, Straus & Giroux, 2011).

Dylan Schmorrow, who likes to cite a *Far Side* cartoon to explain what they'd initially set out to do. In the cartoon, a student in a classroom raises his hand and says, "Mr. Osborne, may I be excused? My brain is full." Schmorrow's original idea was simple: He intended to figure out when and which part of the brain was saturated and then send new information someplace else.

He did so by funding the development of new sensing technologies that could monitor the brain and analyze its signals in real time. The key to augmented cognition, Schmorrow initially believed, was managing the flow of information to different parts of the brain including working memory, that mental blackboard we use to temporarily hold the conscious information we need to act on the world. We have, it turns out, several kinds of working memory, some for spatial information, some for verbal, symbolic information. And when one fills up, it doesn't necessarily mean the other one is out of space, too. Using emerging brain-scanning technologies Schmorrow sought to track these brain states in his pilots in real time, detect when they were becoming overwhelmed, and design interventions that might help the pilots more efficiently make use of the flood of sensory details Potter and others had identified.

In one of Schmorrow's most impressive demonstration projects, one of the teams he funded developed a brain–computer interface that allowed their experimental subjects to fly as many as *twelve* drones simultaneously with hardly any mistakes. The team, led by Boeing, did so by hooking their pilots up to a brain scanner that fed real-time brain data into a pattern recognition program that had been calibrated to detect specific patterns associated with different kinds of information overload. When it detected that certain areas of the brain were maxed out, the computer would change the way information was presented. If the machine detected straight-out cognitive overload, for instance, it might gray out much of the screen in front of the pilot, reducing all distractions and leaving only the areas of the screen relevant to the most urgent task. If the computer detected that the pilot's visual attention was waning, it might announce audibly: "*The screen is changing and you need to pay attention.*" If "verbal" working memory was full, it might reroute information to "spatial" working memory areas, delivering pictoral messages on the screen rather than verbal commands.

Later, Kruse and Schmorrow sought to apply these same brain-scanning technologies to detect flashes of recognition on the very edge of our awareness. One of the efforts they funded was created by a Columbia University engineer named Paul Sajda, who studied the brain's visual system. In the mid-1990s, Sajda had toured the National Photographic Interpretation Center (NPIC), an analysis center established by the CIA in the 1950s, in Washington, D.C. From just a few pixels on the screen in front of them, Sajda noticed, the men and women working there could pick out a satellite dish, or well-camouflaged bunkers amid crowded cities and empty deserts. They seemed, in fact, barely to have to look. It was clearly another demonstration of Klein's pattern matching.

Sajda was surprised to learn, however, that these very same analysts were having trouble keeping up with their work. The problem was there were just too many pictures to look at. Even if the analysts stayed glued to their desks and stared at screens 24/7, they didn't have a chance. They were drowning in a data deluge.

Sajda was aware of Potter's findings on the human visual system and wondered if there might be a way to exploit them using technology.

"Humans have great general object recognition," he reasoned. "Computers are very good at chugging through lots of data." Perhaps, Sajda thought, he could find a way to "marry" the two. Much of the money Schmorrow had poured into research as part of the augmented cognition program had gone into developing non-invasive, portable brain scanners that used technologies like the electroencephalogram (EEG) and a technique called "functional near-infrared spectroscopy" to gather real-time brain data and stream it to a computer. And in fact, the technological advances in these areas funded by AugCog are, most agree, one of the program's greatest legacies.

At the time, these technologies weren't yet up to the task of precisely homing in on the exact brain areas associated with the kind of expertise Cohn would later seek to study. Even so, Sajda wondered if there might be a way to identify a neural signature associated with the recognition of something familiar—during that fleeting moment when pictures appeared and then disappeared from the mental scratchpad of working memory.

Sajda had in mind one particular kind of neural signature he'd read about in the literature. In the 1960s, scientists experimenting with EEG demonstrated that they could reliably detect a particular pattern of neural activity roughly 300 milliseconds after a subject saw a photograph or visual stimulus they recognized or that was anomalous to the other images they had seen. They called this neural signature the "P300 response." Back then, it took multiple trials and hours and hours of tedious postexperiment analysis to find even this vague telltale trace in the data. But if you dug through the data long enough, and cleaned out all the other noise, there was unmistakably something there—a consistent pattern of change in neural activity that you only saw when the subject spotted something their brain was somehow primed to look for. The pioneering scientists who detected it didn't have the technology to identify what exactly was happening in the brain. But that didn't matter.

With DARPA funding from Kruse's program, Sajda invented a device called "cortically coupled computer vision" (C3Vision). Sajda put his subjects in what was essentially a modified shower cap that held sixty or so wires, then he sat them down in front of a screen and flashed a blindingly fast stream of pictures—as quickly as 10 per second. It was way too fast for anyone to actually examine and ponder them. Remarkably, however, Sajda demonstrated it was enough to evoke the P300 response. If a subject saw a picture that he or she had been told to look for—say a satellite dish in an overhead shot thick with houses—and even if it was on the very edge of conscious awareness, even if it went by in a blur, the machine detected the P300 response in real time. Sajda then ran that photograph through a computer program that broke it down by visual characteristics (the location of specific lines, the contrast in a particular area). Next his program chugged through thousands more images, pulled out those that were somehow similar visually, and then sorted them so that the analyst saw them first. Sajda's invention could do all of this in seconds, vastly increasing the likelihood the analyst would see photos of interest during his limited time in front of the screen.

"Our device will do a quick triage and allow analysts to jump from region to region in ways that save time," Sajda says.

Kruse won funding to test out a device on analysts at the National Geo-Spatial Intelligence Agency, and Sajda was able to demonstrate he could help analysts sort through tens of thousands of photographs at speeds that would have been unimaginable before.

In 2007, Kruse and another DARPA program manager launched a second program, this one given the enticing moniker "Luke Skywalker's binoculars," after the high-tech pair of eye gear Luke uses to scan the far horizon in the first Star Wars film. Kruse's aim was to build a device that soldiers could use out in the field as they scanned the horizon for threats. With funding from that program, in 2013, DARPA contractor HRL (a division of Boeing) unveiled a machine capable of spotting moving vehicles 10 kilometers away, and dismounted enemy soldiers at 10 times the previous distance in darkness or in daylight. The device, called the Cognitive Technology Threat Warning System, also allows a soldier to monitor all 360 degrees of the area around him, while vastly expanding the speed at which he can detect threats.

It does so by creating a blindingly fast pastiche of 5-frame-long video clips collected by optical sensors and then flashing them in the center of a soldier's field of vision. A portable EEG setup measures the soldier's real-time brain activity, looking for the P300 response, and pulls out the clips of interest for further review. By rapidly cycling through images taken from all points of the compass, the machine can allow a soldier to keep watch on a field of vision far wider than possible with his own limited eyesight. According to the contractors involved, the device can almost double the ability of test subjects to spot threats.

Cohn was aware of all these efforts. But he wanted to go further. Cohn wasn't interested in building a *device* to enhance performance in the field—he wanted a device or a technique he could use to actually train and hone the human mind, so you wouldn't need a computer to detect a threat in the field.

Cohn also wasn't satisfied with the P300 response, which didn't tell you much about *what* was actually happening and *where* it was happening in the brain.

Cohn realized he would first have to somehow demonstrate he could find a much more localized and specific brain signal. In order to prove it was "intuition," he would then have to prove it was present when his test subjects *knew something without knowing they knew it.*

The idea, however, was not as difficult as it might sound, because there is, in fact, a rich body of scientific literature that provides some powerful clues as to how to test for this and where to look. The literature comes from the study of individuals who are in many ways on the other end of the spectrum of functionality from Klein's heroic firefighters—amnesiacs.

If you want to understand the capacity of the brain to retain information outside conscious awareness, individuals who have lost all ability to form new long-term memories are the perfect patients to study because they can't consciously know they know anything new. How could they? They can't remember learning new things.

By studying amnesiacs in the twentieth century, scientists discovered something that genuinely surprised them—and which to this day seems counterintuitive: that even without the capacity to remember what we had for lunch yesterday, natural selection has imbued all of us with powerful machinery that ensures we can continue to hone those instinctual gut feelings that might, in a pinch, save our lives. Scientists now have a name for this process: "implicit learning."

One of the first recorded accounts of implicit learning in an amnesiac was made more than a hundred years ago by the Swiss psychologist Édouard Claparède. In 1911, he reported the case of a forty-seven-year-old amnesiac with a condition related to thiamine deficiency that had eaten away at an area of her brain known as the hippocampus, a curving, seahorse-shaped structure buried far below the outer cortex. This patient still retained memories formed before her illness—she could name all the European capitals, do arithmetic, and carry on a conversation. But she couldn't recognize the doctors she saw every day, no matter how many times they reintroduced themselves.

Claparède wondered if there were exceptions. So one day he concealed a pin in his palm before shaking the patient's hand. When the pin pricked her skin, she recoiled in pain. The next day, the patient claimed to have no conscious memory of the previous day. And indeed, when she encountered the doctor she behaved as if she had never seen him before. Yet when Claparède extended his hand to introduce himself again, the patient refused to shake it. She could not explain why she looked askance at his outstretched palm. She had no conscious memory of the pinprick from the day before. But somehow, somewhere she knew that shaking his hand was a bad idea. It's not a stretch to say she had an "intuition."

Much of what we know and have learned since about implicit memory—things we know but might not know we know—comes from the study of another amnesiac, one who lost his capacity to form long-term declarative memories after undergoing an experimental brain surgery more than forty years after that pinprick. Henry Gustav Molaison, known to the world as "H.M." until his death in 2008, was just twenty-seven in 1953 when he agreed to undergo a trial procedure in the hopes of relieving the severe epileptic seizures that had tormented him since he fell off a bicycle as a child.

After drilling into Molaison's skull, a Connecticut surgeon named William Scoville sucked out a small section of neural tissue in an area deep in the young man's brain called the "medial temporal lobe," which contains both the hippocampus (the area that had been lost to Claparède's patient) and a nearby, almond-shaped area called the "amygdala." Then the doctor did the same thing on the other side.

Molaison's seizures were reduced dramatically. But the unexpected side effects quickly became apparent: From that day on, Molaison was condemned to live in a state that his biographer Suzanne Corkin, who spent decades alongside the neuroscientist Brenda Milner studying him, used as the title to her 2013 book documenting the experience, *Permanent Present Tense.*

It was through H.M. that scientists first began to understand the essential role of the hippocampus and related brain structures in the formation of long-term memories—which over time would make Molaison one of the most famous and influential scientific subjects of the twentieth century. Before him, many scientists wholly dismissed the idea that there was just one part of the brain responsible for memory and that it could be traced to a specific location. After H.M. the idea that the structures of the medial temporal lobe were essential to forming new memories was no longer disputed. Though Molaison could still hold information briefly in his mind, it was as if the filing clerk responsible for encoding and placing this information in long-term storage had left the building. Once a piece of information disappeared from Molaison's view, to his conscious mind it was gone forever.

Yet the very first time Corkin's mentor, Brenda Milner, met

Molaison, just three years after his surgery, in 1956, she realized this was not the whole picture. A native of Britain, Milner had studied at McGill University under the great psychologist Donald Hebb, whose ideas on "Hebbian learning" and neuroplasticity we learned about in the last chapter. He arranged for her to also work with the neurosurgeon Wilder Penfield. In 1955, Penfield dispatched Milner to Hartford, Connecticut, to examine Molaison after meeting his surgeon at a conference.

Milner gave Molaison a battery of tests, which confirmed his complete inability to form long-term memories. But one test seemed to contradict the rest of the results. On three consecutive days, Milner instructed Molaison to trace a five-pointed star on a piece of paper, keeping his pencil inside the border of the star. To make the task challenging, Milner mounted the paper on a wooden board blocked from Molaison's view. To trace the star, Molaison had to reach around a metal barrier and guide his movements by watching his hand in a mirror placed across from the barrier.

Since the image in the mirror was reversed, it was an awkward task, and a new skill that had to be learned. But as Molaison repeated the task that first day, he began to adjust to the conditions, and the accuracy with which he could trace the star steadily increased. On the second day when he came in, Molaison had no recollection of ever having performed the task before (or of having met Milner). But when Milner readministered the test, Molaison's scores suggested otherwise—they were almost as good as they had been at the end of the day before. By the third day, the task had become so easy that even Molaison was impressed.

"Well, this is strange," Molaison remarked proudly, after nearly flawlessly tracing the star for what he believed to be the first time ever. "I thought that that would be difficult. But it seems as though I've done it quite well."

In the years that followed, scientists would demonstrate that this "nondeclarative" motor memory also extended to other realms. In

1968, Milner decided to test Molaison's ability to unknowingly develop perceptual expertise—and use it to make judgements based on limited perceptual information (something directly relevant to Klein's firefighters and the Spidey sense seen in Cohn's soldiers).

Milner showed Molaison twenty line drawings of common objects and animals, such as an elephant and an umbrella. There were five sets of the same line drawings. But the first set of pictures contained just a few fragments of each object—so few that it would be virtually impossible to identify the object on the first go-round. Each subsequent set offered progressively more fragments, until the final cards displayed the complete, and easily recognizable, object.

"I'm going to show you some pictures that are incomplete," Milner told Molaison. "Tell me what the figure would be if it were completed. Guess if you are not sure."

At the beginning of the first day, Milner showed Molaison the most fragmented cards first to establish a baseline, and tallied his mistakes. He did not get any correct. Then she showed him the next set, with the order of cards mixed up so he could not anticipate them, and told him they were getting a little easier. After going through all five sets—ending with the complete picture of the object—Milner started the process over again. By the fourth trial, Molaison was able to guess the objects depicted on each card with no errors, even when the cards contained only those initial fragments—the very same fragments he'd previously looked at blankly with no recognition whatsoever.

The true test, however, came an hour later, enough of a time lag to ensure Molaison had no explicit memory of ever having taken the test before. When Milner readministered the test, Molaison's scores still improved. On some level, he had retained his ability to correctly classify the fragments, even though he believed he had never seen them before. Corkin returned and readministered the test more than a decade later—and Molaison still did better than the first time.

Researchers have since confirmed the existence of similar implicit, or unconscious, learning in experiments with subjects who have normal memory capacity. In one paradigm, researchers instruct volunteers with normal memory to watch a computer screen for the appearance of a single blinking asterisk in one of four possible positions. Each screen location is assigned a button on the keyboard: A, B, C, or D. Every time the subjects spot an asterisk, they are instructed to press the key associated with its location to show they have seen the asterisk. The test usually consists of a sequence of twelve different asterisks, delivered one after the other, requiring subjects to select one of the four buttons twelve times.

What subjects are not told is that the sequence of twelve spatial locations often repeats itself. The subjects rarely if ever consciously register the fact that the pattern of asterisks is repeating itself. Yet, remarkably, after the patterns have been flashed enough times, their reaction times speed up—implying that the subjects are unconsciously learning the sequences, anticipating the locations that will follow, and moving their fingers faster to the correct keys. (When researchers scramble the sequence, the subjects don't consciously notice the change, but reaction times slow down.)

It's not a stretch to equate this kind of complex, unconscious perceptual learning, which other researchers later expanded in studying H.M. to include other higher-level sensory modalities, to the learned abilities of Klein's expert firefighters: the knowledge they used to perform what he called "pattern matching."

When placed in a situation with limited perceptual information—like a burning house with an apparent kitchen fire that is not loud enough to account for the heat in the room next door—Klein's firefighters just "knew" something was wrong. That fire chief couldn't say how he'd saved the lives of his men. But somehow he knew. The firefighter did not have ESP, as he believed for years afterward. The chief had seen many hundreds of fires over the course of his career; each time he had been exposed to all the perceptual cues that normally went with them. Now when he looked at the conglomeration of limited fragments that day in the kitchen, he could fill in the blanks.

Somewhere outside his conscious mind, some part of him did pattern matching and reached an urgent conclusion: Danger—get out now!

In order to win funding to study intuition Joseph Cohn had to somehow reach inside the brain and prove he could see intuition operating in real time. To help design experiments, Cohn recruited a couple of intuition consultants who had previously worked with Gary Klein. He also enlisted a team of researchers from the University of Oregon* led by neuroscientist Don Tucker and his colleague Phan Luu, two researchers who had helped design sensing technologies for the augmented cognition program years before.

Tucker and Luu were interested in how things we see with our sensory systems interact with the more primitive and emotional part of the brain called the "limbic system." It was a promising place to look for instinct. Some theorists argued that the limbic system developed in evolutionary terms to manage the fight-or-flight instinct. It thus stood to reason that if instinctual alarm bells are ringing—if one's hair is standing on end and something "is wrong"—it's the limbic system that is doing the ringing. Tucker and Luu hoped to somehow capture and characterize that moment when anomalies detected by the sensory areas of the brain pulled the limbic system into action, entirely outside conscious awareness. They wanted to measure gut feelings.

---

* Tucker and Luu used fMRI scanners to get a detailed map of the geography of each subject's brain, then synched that data up with a 256-channel EEG skullcap, which offered superior ability to monitor brain spikes in real time than the fMRI, but less accurate spatial detail.

Tucker and Luu decided to show test subjects pictures of incomplete objects, fragments similar to those shown to the amnesiac Henry Molaison by Milner in 1968. But in this experiment, Cohn's team would also use state-of-the-art brain-scanning technologies far more advanced than anything that existed back in the 1960s and 1970s to find a neural signal that was present in specific areas of the brain only when their test subjects subconsciously recognized that there was a complete object there. Could Tucker and Luu show that they could tell, just by looking at brain scans, whether someone *knew something without knowing they knew it*?

Tucker and his team scanned the brains of twenty-two students using fMRI and a dense array EEG (dEEG) as the subjects sat in front of a computer monitor and looked at two hundred different images flashing at an extremely rapid clip (less than half a second) before their eyes. One hundred and fifty of these images contained fragments that were parts of actual objects, just like Molaison had viewed in that seminal Milner study. A pixillated picture of a bed, for instance, or a cup.

Tucker and Luu had removed so many blocks of pixels, however, that it was virtually impossible to know at such a rapid speed for sure what the picture was. For comparison, the team also added in fifty images composed of nonsense fragments, pixels chosen at random by a computer and scrambled. They did not represent fragments of actual objects, just visual "noise."

The instructions were simple: Guess, relying on your best impressions, whether an image contains an object, or whether it is just random pixels. You don't have to name the object—in fact, even if you try, you probably won't be able to. Just take your best guess as to whether something is there.

"Just tell us in your gut when you feel that there are things to be observed in the scene or not" is how Cohn put it.

Sure enough, the participants correctly intuited there was an object hidden in the fragments about 65 percent of the time. They incorrectly guessed there was an object hidden in the nonsense fragments about 14 percent of the time. But most important, the researchers spotted a neural signature that allowed them to discern, just by looking at their brain scans, whether their subjects had gotten it right.

In the cases where the subjects guessed correctly, brain activity began to diverge from that recorded in the incorrect guesses 100 milliseconds before the impression seemed to enter the conscious mind, about the time it takes to blink an eye. Sure enough, the brain activity consisted of a flash of rhythmic electrical oscillations originating in the sensory areas of the brain and leading to the limbic system, the primitive subconscious areas that constitute the seat of emotion and the fight-or-flight response.

At the same time, the brain also began to generate a second repeating and more global pattern of brain waves, oscillating at what appeared to be a frequency called the "theta band." The rhythm is often seen when the brain is mobilizing disparate areas into an ad hoc network, says Cohn. It's like the beating of a drum in an army unit, marshaling the troops to begin to step in sync, but here in preparation for more cognitive analysis.

"What that told us was that if you were having an intuition, A, your limbic system was activated, which is why you have that gut feeling, 'Wow, something's going on!'" Cohn says. "But, B, other parts of the brain start to get pulled in to help you make sense of that information, and that's what the pattern of neural activity told us."

Tucker and Luu's study did indeed seem to suggest that a firefighter whose senses detect that he might be in mortal terror gets "that feeling" because a signal has arrived in the areas of the brain we rely on for our very survival—and arrived there long before it was sent to the conscious areas of the brain. With incomplete information, this area begins to prepare the body for a response should it be needed once the conscious mind has enough sensory information to make an informed judgment.

Soon after Tucker and Luu's study, Cohn moved from DARPA to the Office of Naval Research. Despite the change in venue, he was intent on taking the next step. Sure enough, within a year he had gained approval for a much larger, four-year, $3.85 million program not just to continue to characterize intuition, but to begin to find ways to train it.

In 1984, researchers at the University of Pennsylvania working with amnesiacs noticed that after a lengthy exposure to a particular word or concept, such as dogs or cats, their patients would soon forget they had the conversation. But if the researchers then asked the patients to initiate a new conversation on any topic of interest, the amnesiac patients would often start to talk about dogs or terriers or the greyhound racetrack, or cats or Siamese kittens or *Garfield*. They'd forgotten the initial conversation but the brain remained somehow "primed" to talk about dogs or cats.

If you hear a jingle in a television advertisement for the first time, and then, ten minutes later, suddenly find yourself singing it, that's "priming." Advertisers and political candidates use priming to get in our heads.

Priming appears to be a powerful and ubiquitous form of implicit memory, discovered relatively recently only because it is entirely unconscious and thus hard to detect. We don't know we have been primed unless someone points it out. Yet priming is so crucial to the human experience that its neural correlates can be found all over the brain, and it is present in all of us. The priming effect is just as powerful in elderly subjects suffering from long-term memory deficits as it is in young adults. What's more, researchers have even found that priming effects are as strong in a three-year-old as they are in college students. Even alcoholics, who can go into blackouts and forget what happened the night before, might still wake up singing a song from the previous evening, due to priming effects. All of these groups are just as likely to spontaneously launch into a conversation about dogs without knowing why, if you prime them to do so.

Over the last twenty years, a number of researchers have used the latest brain-scanning technologies to try to understand exactly what happens when priming occurs. From that they have expanded the studies to look at other forms of implicit memory. This provides a starting point, a neural signature brain scientists and psychologists can use to try to understand what exactly happens in the brain when we gain new implicit knowledge and begin to develop those templates that Cohn is so interested in finding ways to train.

The results are remarkably consistent. When subjects see an object or hear words or are exposed to some sort of pattern for the first time, one of the first places one finds evidence of the stimuli is in the sensory-processing areas of the cortex. If it is a visual stimulus, you'll see activity in areas of the visual cortex; if it is a sound, it will show up in the auditory-processing areas of the brain. If it is a complex idea, the prefrontal cortex might alight.

But here's the thing: When subjects see that object or word or pattern the second time, the same areas also activate—but the activity in those areas, with each repetition, is likely to be *reduced*. This phenomenon is called "repetition suppression." And though it might at first seem counterintuitive, its cause is, in fact, easy enough to explain. The brain is simply growing more efficient at processing the signal. It is doing so because of neuroplasticity.

"Every area of the cortex has the adaptive ability to rewire itself based on experience," explains Paul Reber, an expert on implicit learning at Northwestern University. "So if you are in a situation where there are elements of the environment that are familiar and fall into a known pattern, or follow learned contingencies, one signature we might look for is efficient processing of information about our environment."

Reber is leading the current phase of the intuition project, and is looking for ways to shape and train the mind and mold the filters that help us process the information around us. He has been studying priming, implicit learning, and their neural correlates for decades.

The more often we see something or hear something or experience something we have already experienced before, the fewer neurons will activate in the sensory areas of the brain—yet the more robustly they will do so. This makes sense, Reber explains, if you consider how neuroplasticity works. The cell membranes of all neurons have a faint electrical charge, which subtly changes in response to every chemical signal a neuron receives from other neurons at its synapses. It's only when a neuron's electrical charge is pushed up past a certain threshold that it will fire an electrical spike of its own. As we learned in the last chapter, every time two neurons fire together their connections to one another strengthen—in other words, the electrical signals passing back and forth between the two neurons each time one of them fires grow more powerful. Conversely, every time two neurons fire apart, those signals slightly weaken.

If you think of it that way, it makes sense that the more times we see a particular object, the more efficient the connections that represent it in our visual cortex become. With each exposure, some connections will become stronger—because two neurons have fired together and are thus wired together—and some will become weaker, because they have fired apart and thus are wired apart. Through this process of Hebbian learning, a new circuit is sculpted by experience. Over time, fewer neurons fire, but those that do fire together grow more connected and sensitive to one another. The brain becomes more efficient. Fewer neurons fire, and those that do, fire far more robustly.

This simple rule, Reber realized, has powerful implications for training.

Reber argues there's a "statistical" element to shaping the connections of the brain, and thus to developing the "filters" that might allow a trained soldier to subconsciously pick up on visual cues others might not see.

"Every kind of synapse in the brain has some inherent plastic capacity," Reber says. "So one thing to understand about the mechanisms behind implicit learning is it picks up whatever statistics you're exposed to."

The more you have been exposed to a certain cue, or group of cues together, the more likely you are to respond to those cues in the future when you see them again, and the more likely the areas of the brain that encode images you have come to associate with those cues will also become active.

From this a practical logic for training intuition emerges. Reber believes the best way to train it is with repetitive drills of the same sort one might rely on to develop motor skills, like serving a tennis ball, riding a bicycle, or swinging a baseball bat with the proper form.

Reber, who began his career studying amnesiacs, has spent the last thirty years studying how this process works. This interest makes him an ideal expert to help Cohn and his successor Peter Squire take the study of intuition to the next level, because if you understand how these unconscious filters develop, then you can come up with ways to train for them. You can take that kid from Wyoming and design a program that will allow him to intuitively sense when he is in danger in Iraq.

A couple of years ago, Cohn handed the intuition program over to Squire, who is overseeing an effort that is proceeding along three fronts.

The first task is to translate laboratory work of the kind Reber has pursued in more realistic settings relevant to soldiers. Over the years, Reber has found similarities between what happens in the brain when we improve at a physical skill, such as riding a bicycle, and what happens when we become better at a visual recognition task. While activity becomes more efficient and involves a smaller number of neurons in the visual cortex—"repetition suppression"—he has found that it actually *increases* in another area of the brain, called the "basal ganglia." The basal ganglia has previously been shown to play a key role in the learning of complex motor tasks, like riding a bicycle or bouncing a basketball. But the idea that the seat of so-called motor memory also plays a role in expediting the speed with which we can process visual information is a newer one.

The involvement of the basil ganglia provides further evidence to Reber that repetition is the best way to take that kid from Wyoming and train him to be able to sense the presence of IEDs in Iraq or Afghanistan without consciously thinking about it.

"If you want to build up this implicit learning layer you probably actually have to do something that looks a bit more like drill training. So build lots and lots of scenarios, run somebody through a few hundred scenarios where the statistics can be embedded in what they're practicing going through," Reber says.

Eventually, even before the soldier is consciously aware of it, neurons in the visual cortex will automatically respond to evidence of freshly dug dirt, if it is also in the presence of another cue like a stray piece of wire.

To prove this has utility for the field, Reber and his team are now attempting to demonstrate they can get the same results in a situation more akin to the chaos of the real world. Squire has asked them to develop testing in the context of a simulated danger zone used to train individual marines or units operating together in "virtual space."

"Some of the terrain might look like Afghanistan, for instance," says Squire. "We want to see if we can elicit the same intuition effects—and what kinds of modifications and trainings we might design to accelerate it. It could be IEDs, or the presence of snipers or terrorists. There are certain regular patterns that might be disrupted."

Those patterns might be something as subtle as discoloration in the dirt, or a lack of normal activity on the street.

Reber has also begun to examine how one might train soldiers to recognize and know when to pay attention to "gut feelings."

"Conscious processing and implicit learning don't always play nicely together," Reber says. "When you focus on one, it can hide the other."

So as another part of the study, Reber is collaborating with his colleague Mark Beeman. One question they are asking is what exactly happens in the brain when we experience a moment of insight, that "aha" moment when something we don't know we know becomes conscious.

In one test, Beeman and his team gave subjects in the laboratory three words, such as *crab, pine,* and *sauce,* and asked them what word might go with all three.

"People typically try to do it by explicitly coming up with every associate they can," Reber says. "But Mark has shown they are unlikely to solve it this way. People can struggle a long time. They can't quite see the answer. But what they'll experience on some percentage of trials is a sudden insight. . . . They're struggling, struggling, struggling, and then suddenly they think, 'Oh, apple! It's apple!'"

Though the subjects invariably report that the answer seems to strike them from out of the blue, Reber and Beeman are convinced that these sudden insights are in fact the result of implicit memory processing.

"It's activity spreading among related concepts that's happening outside awareness," Reber says.

By analyzing the neural activation patterns of the moments before, during, and after insight, Reber and Beeman have discovered something interesting: The areas of the brain that are active at the moment of insight—and, perhaps just as important, inactive—are similar to the areas that are active and at rest when subjects successfully perform a wholly different kind of task.

In this second task, the two neuroscientists showed pictures of oversize letters made up of clusters of different smaller letters—a huge *H,* for instance, composed of scores of tiny *t*'s. They instructed the participants to name the larger letter. To get it right, the subjects had to "step back" and look at the overall picture. They had to, in other words, relax their focus on details and force themselves to "see the forest for the trees."

"When you have to do that, there's a part of the anterior frontal lobe that's active, and that same area is also active right before you suddenly have this insight and say, 'Oh, wait a minute—it's apple!'" Reber says.

Reber hopes to find and demonstrate ways to promote insight, and train people to be more open to intuition by teaching them to recognize and ease into this mind state.

"What does this mean if you're out in the real world making decisions? What do you say to a firefighter?" he asks. "Well, maybe you have to say, 'Don't focus on the perceptual details.' Maybe you have to kind of step back from an intense local focus on perceptual detail and think more globally about the overall circumstance, to let the intuitive information percolate up to your awareness."

There's a certain intuitive logic in much of what Cohn's researchers are finding. It makes sense to me that visual drill training is the way to hone the ability to recognize a threat. And, in fact, Reber's suggestion that individuals might be trained to recognize particular brain states associated with better performance relates back to some of the earlier work that grew out of the augmented cognition program.

Amy Kruse also wondered if you could tell whether someone was an expert at a task just by looking at their brain—and whether it might be possible to track the changes as someone progressed from a state of beginner to a state of mastery.

To answer these questions, Kruse funded Chris Berka, the CEO of a company called Advanced Brain Monitoring (ABM), to perform a series of fascinating experiments as part of a project known as the Accelerated Learning Program. ABM, located in Carlsbad, California, recruited the top-level marine sniper instructors from nearby Camp Pendleton. Every time a volunteer arrived for testing, Berka and her team placed a dense twenty-four-channel array of electrodes over his skull, wired him up to track respiration and heart response, and then handed the volunteer a jury-rigged M4 rifle capable of taking precise measurements that included muzzle wobble and trigger squeeze. Then Berka and her team had the expert marksmen go through the process they normally went through before squeezing off a shot at a target, and looked for interesting patterns in the data. When she and her team tallied up the results, they discovered something fascinating.

"What we found is that in the two to five seconds preceding a perfect shot we saw the exact same psychophysiological profile," Berka says.

The heart rate slowed down, followed by a long, slow inhale, and then an exhale that coincided with the shot (snipers are trained to exhale on the shot). But even more interesting was the presence of two distinct EEG signatures. The first was an increase in what's known as "midline theta" activity, a train of rhythmic waveforms in the theta range (4 to 7 Hz, the number of brain wave cycles per second). Simultaneously there was a burst of "alpha" activity over the left temporoparietal region of the brain, a specific pattern of neural oscillations between frequencies of 7.5 and 12.5 Hz, thought to be associated with synchronous activity.

The midline theta seemed to be a neural signature that reflected the sniper going through mental checklists, or visualizing the perfect shot. The sniper's increase in alpha activity, Berka notes, has been documented in a number of other laboratory settings and is a well-known signature of focused attention.

"So, you're kind of shutting down the incoming sensory information, focusing just on the target, and getting a perfect shot," she says.

"Those were the three signatures," she says. "Heart rate deceleration, theta midline, and alpha over the left temporoparietal region. And what was fascinating is not only did we see that in thirteen coaches—shot-to-shot exactly the same pattern—but if we just had them sit in the room with all the gear on and imagine shooting, we saw exactly the same pattern."

When asked, the coaches reported that they were in fact subjectively aware of the feeling associated with this neural pattern.

"Yes, I know that I'm in that state," Berka recalls more than one sniper telling her. "I know it when I hit that point where I'm going to take a perfect shot. And I use that even in combat situations."

Many told her that the state was accessible at will—and that they had learned how to access it on their own over time as a by-product of experience. It was, some said, "almost like a little switch." "It doesn't really matter what's going on around me. I'm able to completely focus on taking the shot," Berka recalls several of them telling her.

After identifying these neural signatures, Berka and her team recruited 150 civilians and 150 marines with various levels of marksmanship skill, all nonexpert. Then they developed a prototyped video introduction to marksmanship to ensure consistency of training and divided the volunteers into two groups. The control group was allowed to practice shooting on their own and view additional instructional videos. The experimental group was instead set up with EEG and heart rate monitors and provided with real-time visual or auditory feedback so the subjects could monitor when their heart rate and brain waves moved in and out of the ideal expert states detected in the best marksmen. (Most people preferred haptic feedback, which came in the form of a buzzer clipped to their shirts that vibrated in accord with their heart rates until they achieved the optimum brain state for shooting, at which time the vibrations ceased.)

"First we just trained people without the rifle to move as much as they could towards the expert state," Berka says. "Then you leave the system on and now you pick up the rifle and you shoot."

The feedback and expert state accelerated the marksmanship skill training by a factor of 2.3. In other words, those with the feedback learned to become master marksmen twice as fast, regardless of innate skill. The feedback, Berka says, allows individual to recognize the "expert" brain state and learn to control it.

This idea that we have distinct "brain states" that we can be trained to recognize—whether they are associated with hitting a bull's-eye, or spotting danger out in the field—is a familiar concept. Athletes often talk about being "in the zone," swishing the basketball with ease, seeing the stitches on a baseball. I've experienced this state of mindless ease and total immersion while jogging, or reading, writing, or playing music. It strikes me that there is a specific brain state associated with focus and the zone and I do recognize it when I enter it. The world does go quiet. I am focused. Things often flow.

Thus to me, the use of brain-scanning technologies to train these kinds of skills does indeed seem likely to become more prevalent in the years ahead. That raises an interesting question: If we can use brain-scanning technologies to detect states associated with implicit learning, instinct, and expertise, what else might we be able to detect? How far might we push it? And what might it allow us to do in the medical realm?

CHAPTER 6

THE TELEPATHY TECHNICIAN

DECODING THE BRAIN AND IMAGINED SPEECH

It took trips to three different neurologists before David Jayne finally found one who could explain the cause of his mysterious symptoms. Why it was, for instance, that an athletic, twenty-six-year-old man, in the prime of his life—six foot three and two hundred pounds, thank you very much—kept dropping ketchup bottles as if they were five-hundred-pound weights. Why his left triceps kept twitching. Why he suddenly found himself unable to perform even the most basic manipulations of finger and thumb needed to pack and tie spun deer hair onto the ends of fly-fishing hooks.

It was this final indignity that Jayne found most irritating. Back then, Jayne was the very picture of robust and youthful vitality, a sportsman with big blue eyes, "handsome as hell," recalls his sister Sue Ann Cecere. David had been senior class president in high school, and "pretty much the boy about campus" at the University of Georgia. "He had a good time and then he got serious," Sue Ann says. After graduating, David married his college sweetheart, Melissa, and landed on the fast track in the Domino's Pizza corporation, jet-setting around the Southeast opening franchises.

But when given the choice, there was never any doubt where David most wanted to be. He was a born fly fisherman, at home waist deep in a fast-moving Georgia creek, casting softly into a run of steelhead trout. David had grown impatient with the nagging medical mystery that he was almost certain would amount to nothing more than a pinched nerve. He felt the powerful call of Atlanta's Chattahoochee River just a few miles from his office on that beautiful spring day of the doctor's appointment that would change his life.

"I always had my fly rods, waders, and float tube in the back of my truck," Jayne would later recall. "The Chattahoochee River was to my left and I proceeded to turn right."

In the waiting room at Emory University Medical Center, Jayne noticed an attractive young woman seated between her parents. "My attention was drawn to her because her body had the appearance of an extremely well-loved rag doll," he recalls. What could be wrong with her—why was she so limp? He wondered. Then his name was called.

Years later, Jayne would remember with a visceral clarity the moments that followed: The paper sheet on the "rock-hard" examination table crinkling beneath him when he sat down. The late afternoon sunlight streaming in through a tall narrow window to his right. The way the neurologist clutched his thick files to his chest "like a schoolboy carrying books" as he solemnly entered the room and took a seat on a low stool in front of him. And, finally, the matter-of-fact tone of voice as he uttered the following words: "Mr. Jayne, it is amyotrophic lateral sclerosis, ALS. You might have heard it referred to as Lou Gehrig's disease."

The doctor wasn't talking about that "rag doll" in the waiting room—though she had it, too. Jayne felt numb as the doctor described what lay ahead for him: "How I would become totally paralyzed, losing the ability to speak and eat, and how I would eventually suffocate," Jayne recalls in his memoir.

Then there was more news. The same week David Jayne's doctor told him he had at most three to five years to live, a different doctor informed Melissa she was pregnant with their first child. It was overwhelming.

"The one thing I desired most in life was to be a father," Jayne says. "I wanted a large family. I had so much I wanted to teach and share with my children." But, he thought bitterly, "I will not even be a memory in my child's life."

David Jayne was not yet ready to check out. But the alternative was equally terrifying. Even if doctors could find a way to help him eat and breathe once his body failed him, he would no longer be able to move, or to speak. He would be left totally "locked in," trapped in a body that no longer worked.

What kind of a life was that?

Few diseases are as frightening to contemplate as those that leave us fully conscious and aware, yet wholly and irreversibly unable to move or speak. That's why so many people stricken with ALS order doctors not to resuscitate them when their lungs finally fail. No one questions this decision. But David Jayne would ultimately choose a different course. He was too young to die, so he resolved that when his disease later began to close in on him, he would fight, and risk facing a nightmare that's chilling even to consider: becoming fully locked in.

"It seems to me," one of David Jayne's neurologists told me of the experience of becoming fully locked in, "that it is something akin to being buried alive."

When miners are trapped underground, we have no choice but to try to dig them out. It is no different for diseases like ALS. In recent years, a small band of scientists have devoted themselves to achieving what amounts to a scientific Hail Mary: They are attempting to harness modern neuroscience and computing technologies and develop new ways to reach into the intact minds of patients like David Jayne. Then these scientists are attempting to access the words these patients are no longer able to articulate using their muscles, and somehow translate them back into language. They are attempting, in other words, to "dig out" these patients by restoring to them not only the ability to move, a challenge that is itself hard to fathom, but to speak. To reanimate their broken bodies.

Restoring speech is a technical feat beyond anything we have yet considered. And the stakes are, arguably, far higher.

Certainly it would be hard to find a scientific challenge more daunting. The three-pound gelatinous mass we call the human brain is never outputting just a single signal at a time. Across the entire surface of the cortex, and down into the brain's very core, at any given moment billions of individual neurons are passing chemical signals back and forth and converting the sum of those signals into hundreds of millions of faint electrical pulses. A powerful computer is the only possible tool that offers a chance of extracting meaning out of this mass of information.

But this challenge is many steps beyond the kind of brain sensing that Joseph Cohn and Peter Squire are deploying to try to capture that flash of magic associated with intuition—many steps beyond the technologies that scientists used to discern activity in Pat Fletcher's visual cortex.

Even if you find a way to "plug" a mushy, living, functioning human brain directly into a computer made of hard plastic and silicone, one capable of translating the signals into the English language, there's no *single* signal to try to capture. To detect someone's thoughts, to listen in on his or her private, unspoken internal monologue and pluck out speech, some argue, you'd have to monitor *millions* of signals at the same time, and figure out what they all mean together. And to do that you'd have to sort out all the irrelevant noise that has nothing to do with imagined words and ideas—the brain signals controlling breathing, for instance, or eye blinks or the feeling that "something isn't quite right" and the floor is about to collapse.

This is a mathematical and decoding challenge that makes Alan Turing's legendary effort to break the secret Nazi code of Enigma during World War II appear almost rudimentary by comparison.

Yet on a frigid afternoon in February 2015, I'm sitting in a hospital room in downtown Albany, New York, as a team of white-jacketed technicians bustle about the bed of a forty-year-old single mother from Schenectady, named Cathy. And they are getting ready to demonstrate that, in fact, it might soon be possible to do just that.

I have been led here by Gerwin Schalk, a gregarious, Austrian-born neuroscientist who looks a little like Liam Neeson might look if the actor were a few inches shorter, a couple of pounds heavier, and spent most of his time stationed in front of a computer monitor. Schalk and I have been waiting several months for someone like Cathy to come along so that he can show me just how far he and other neurological code breakers have traveled since that day decades ago when Hubel and Wiesel first listened in on those neurons firing in a cat's visual cortex.

Cathy is epileptic and plans to undergo brain surgery to try to remove the portion of her brain that is the source of her seizures. To find it, three days ago, doctors lifted off the top of Cathy's skull and placed 117 tiny electrodes directly onto the right surface of her naked cortex so they could monitor her brain activity and map the target area. While she waits, she has volunteered to participate in Schalk's research.

Now, next to my chair, Cathy is propped up in a motorized bed beneath a thin blanket and a wrinkled white sheet. She's wearing a hospital gown, and her fashionable reading glasses are perched precariously on her small, delicate nose. It's hard not to stare at her unusual headgear. The top of Cathy's head, from her ears and forehead upward, is swathed in a stiff, plasterlike mold of bandages and surgical tape. A thick jumble of mesh-covered wires protrudes from the opening at the top of her skull, as if she were related to Jabba the Hutt's sand-colored majordomo, the tail-headed Bib Fortuna.

Cathy's appendage of wires flops over the back of her hospital bed, drops down to the ground, and snakes over to a cart holding $250,000 worth of boxes, amplifiers, splitters, and computers, all controlled by a white-jacketed attendant standing in front of a large screen.

The attendant gives a signal, and Cathy focuses on a monitor sitting on the table in front of her as a series of single words emanate in a female monotone from a pair of nearby speakers.

"Spoon . . ."

"Python . . ."

"Battlefield . . ."

"Cowboys . . ."

"Telephone . . ."

"Swimming . . ."

After each word, a colored plus sign appears on Cathy's monitor and flashes turquoise. This flash is Cathy's cue to repeat each word silently in her head. Cathy's face is inscrutable. There is no visible way to tell what she is thinking. But as she imagines each word, the 117 electrodes sitting atop her cortex record the unique combination of electrical activity emanating from hundreds of millions of individual neurons in an area of her brain called the "temporal lobe." Those patterns shoot through the wires, into a box that amplifies them, and then into the computer, where they are represented in the peaks and valleys of stacked horizontal lines scrolling across the screen in front of the technician. Buried somewhere in that mass of squiggly lines, so thick and impenetrable it resembles a handful of hair pulled taut with a brush, is a logical pattern, a code that can be read if one understands the mysterious language of the brain.

Later, Schalk's team at the Wadsworth Center, a public health laboratory of the New York State Department of Health located in Albany, along with collaborators at the University of California, Berkeley, will pore over the data. Each one of Cathy's electrodes records the status of roughly 1 million neurons, roughly 10 times a second, creating a dizzying blizzard of numbers and combinations and possible meanings (600 million signals a minute).

Yet now Schalk fixes me with an unwavering gaze from across the room and tells me that he and his team can solve the puzzle and, using modern computing power, extract from that mass of data the words that Cathy has imagined. They are quite certain, he promises to show me, that he has begun to figure out how to "read" Cathy's mind.

Gerwin Schalk is not your typical scientist. Until just a few years ago, he'd never read a scientific journal. He knew next to nothing about the human brain. Certainly he'd never met anyone with ALS like David Jayne. Schalk has always had only one passion: computers. Scientists need people like Schalk to have any chance at cracking nature's most complex code.

A native of the medieval Austrian city of Graz, Schalk fell in love with all things digital the moment his father came home with a Commodore Amiga computer when Gerwin was twelve. Schalk spent much of his teen years poring over English-language computer manuals, trying, like an archeologist decoding an ancient scroll, to pull meaning out of the jumble of incomprehensible words, gradually coming to speak the language.

In high school, Schalk used to surreptitiously unlock his school's computer lab windows during the day so he could sneak back in and use the computers at night. At fifteen he gave himself a tag, "MAD," and programmed those computers to flash "MAD is AWESOME" every time someone booted them up. Reversing this trick was beyond the abilities of Schalk's computer teacher, so there it remained—a daily proclamation of Schalk's anonymous awesomeness to greet him in that computer lab for the remainder of his high school career.

At Graz Technical University, Schalk entered a dual bachelor-master's program in electrical engineering and computer science, a notorious crucible that took almost eight years to complete. When he finally finished, he was itching for adventure. So in 1997, when Schalk heard about an opportunity to travel to the United States for a thesis project in a place called Albany, New York, he grabbed for it. Schalk had never heard of the specific field in question: brain–computer interfaces (BCI). But he didn't particularly care what the topic was.

"I thought maybe I would be working on a three-dimensional reconstruction of the jaw from, like, computer tomography images or something," Schalk says. "But it said New York, so I thought I would be near New York City."

Neither assumption turned out to be true. But Schalk's choice would prove auspicious professionally. The job had been made available to students because one of Schalk's local professors was collaborating with Jonathan Wolpaw, a research physician at the Wadsworth Center. Wolpaw would become the ideal guide into a burgeoning field that, it turns out, was ideally suited for Schalk's talents.

Wolpaw was a board-certified neurologist who worked with locked-in patients and was attempting to harness existing technologies to help restore their ability to communicate. A seasoned clinician, Wolpaw knew as much as one can hope to know about the functioning of the brain and the myriad ways things can go wrong. And he was more than happy to share his knowledge.

What Wolpaw needed was someone who could help him make sense of the dizzying amount of data he was able to extract from inside that organ with the rapidly expanding tools of technology. He was, it would turn out, a generous mentor. Within just a few years,

Schalk had not only won Wolpaw's confidence, he had also earned a second master's and a Ph.D. at nearby Rensselaer Polytechnic Institute, graduating with a 4.0 grade point average and the highest score ever recorded in the electrical engineering department's doctoral qualifying exam. And he had snagged a $1.4 million NIH grant and designed a general purpose software for brain–computer interfaces that is now the industry standard and is used by three thousand people worldwide.

But perhaps most important, under Wolpaw's tutelage Schalk had received a first-class education in a field he was to adopt as his own: an obscure-sounding area of inquiry he'd mistakenly assumed was related to three-dimensional reconstruction of the jaw—the field focused on developing brain–computer interfaces, what insiders simply call BCI.

In 1969, another young researcher, Eberhard Fetz, pulled off a feat so far ahead of its time, many in the field of neuroscience huffed with skepticism. Fetz trained a monkey to move a meter arm using only his brain waves.

Fetz wasn't exactly claiming telekinesis, though you might have thought he was from some of the vitriol he encountered. But hooking a monkey's brain up to a mechanical device was a trick not many in neuroscience at the time would have thought to try. And Fetz's path to this out-of-the-box approach was highly unusual.

Fetz had been working toward a physics Ph.D. at MIT in statistical mechanics when he had a life-changing epiphany while tripping on mescaline or possibly LSD (he can't recall) one night. Dazzled by the kaleidoscopic shifts in consciousness he glimpsed that night, Fetz realized he was far more interested in the mysteries of the mind than "chasing the magnetic moment of a particle." During postdoctoral research with a neuroscientist at the University of Washington, Fetz learned to master the same single-neuron recording techniques that Hubel and Wiesel had used so effectively to record the activity of neurons in the visual cortex of cats. But while Hubel and Wiesel had focused on measuring the incoming signals headed to the visual cortex, Fetz's lab had a different target. They hooked electrodes up to neurons in the motor cortex, tracking activity headed out of the brain.

Fetz had the creativity and the technical prowess to go beyond simply taking measurements. To do so, Fetz placed his monkey in a booth facing an applesauce spigot and a mechanical device that moved a meter arm every time it sensed the neuron firing. Then Fetz increased the applesauce reward every time the monkey caused the meter arm to move.

"As he got the hang of it, he increased the rate that neuron is firing pretty fast and intentionally," Fetz recalls. "It was the first demonstration that the monkey could control the displacement of the meter arm by neural activity."

It was a "real hero experiment," Schalk says decades later, shaking his head in admiration. "In 1971 to create a device that would record signals at a high enough resolution, would process it in real time, and would provide feedback in real time was something that really was technically very challenging."

It was challenging enough to eventually win Fetz a high-profile article in the prestigious journal *Science* and establish him as a role model and founding father for future generations of BCI pioneers, like Schalk. But it took a couple of decades for the rest of the field to catch up.

It wasn't until the early 1980s, in Schalk's estimation, that the field received the next crucial piece of the puzzle needed to take BCI to the next step. That's when a young Johns Hopkins researcher named Apostolos Georgopoulos recorded the activity of neurons in the higher-level processing areas of the motor cortex and demonstrated something remarkable: Certain neurons in the motor cortex were preferentially sensitive to specific directions in physical movement and could be used to predict the intended movement of *an entire limb.* Just as those initial neurons tracked by Hubel and Wiesel decades before responded to specific angles of light, Georgopoulos identified neurons that responded to specific kinds of movement—such as a flick of the wrist to the right, or a downward thrust with the arm. What made Georgopoulos's discovery so important was not just his demonstration that you could record these signals milliseconds before an actual movement and predict those movements before they happened—some of these neuronal firing patterns actually guided the behavior of scores of lower-level neurons working together to move the individual muscles they controlled in concert.

These high-level homing beacon–like signals, when analyzed in conjunction with enough other neurons, could be used to extract a remarkable amount of information about intended limb motion.

"Every cell has a preferred direction, and it's the sum of these preferences that determines which way the animal moves," Georgopoulos, now at the University of Minnesota, explains.

Georgopoulos demonstrated that using 240 electrodes, he could precisely predict the direction in which a monkey moved his joystick. A few years later, he demonstrated that he could do the same in three-dimensional space with 570 electrodes. Further, Georgopoulos demonstrated not just that it was possible to determine the direction—he also could tell the velocity of movement and how it changed over time.

This discovery, when combined with Fetz's simple demonstration, had profound implications for those hoping to help the paralyzed. The neurons of the motor cortex tell our muscles to move by transmitting the electrical impulses down the body's equivalent of long-distance telephone wires, bundles of nerves that protrude out of the skull and run down the spine and out to the body's extremities, where they connect and interface directly with the muscles, and cause them to expand or contract.

When those nerve connections are severed—if, for instance, someone breaks his or her spine—the result is paralysis. When the motor neurons that convey the signals out of the brain to the limbs die, as happens in the case of ALS, the person becomes locked in. But in many paralyzed patients the neurons in the motor control centers of the brain itself remain intact, continuing to crackle with electrical activity that flows to where the nerves were severed or died, where the energy is then expelled into the void like a broken electrical wire convulsing with sparks on the sidewalk after a storm.

Paul Bach-y-Rita's famous maxim—"We see with the brain, not with the eyes"—could be said to apply to movement as well. Many quadriplegics and locked-in patients can still "move with the brain"; they can command their arms to reach out and hug a loved one, even if that signal never arrives at its intended destination. They can tell their legs to propel them forward into a standing position that can bear their weight, even if they remain motionless. They can command their lips to move and their vocal cords to sing. When these individuals do think a command, electrical impulses pulse through the motor cortex in waves of coordinated signals that can be detected with the right device.

The idea that these signals remain trapped, yet conceivably accessible, suggests scientists might one day vastly improve the options available for paralyzed individuals and amputees. Even with lifelike, biomechanically accurate bionic limbs, there are still many things amputees, people like MIT's Hugh Herr, for instance, can't do—like stand on his tiptoes to pull a jar out from the back of a cabinet or flex his feet to put on shoes. Or dance with a daughter at her wedding. Even Herr's state-of-the-art limbs rely on preprogrammed algorithms that move the mechanical prosthetic feet only in relation to the movements of his upper thighs. This denies him the ability to spontaneously flex on a whim, or spin around in an eccentric manner just by thinking.

For those who worked with patients who were "locked in,"

what Georgopoulos called his neurons' "tuning curve" suggested something else: that the neurons controlling the muscles in our lips, throat, and tongue, which we use to speak, could also be recorded and that their patterns of activity could be decoded, too. That speech, in other words, could be restored through the use of a synthesized voice.

As soon as bioengineers began to experiment with implantable neural electrodes and attempted to hook them up to actual external devices, however, they discovered a wide array of new challenges. Implanted brain electrodes often move and jar loose, and because of neuroplasticity the populations of neurons that control any given movement or action are also bound to shift eventually. Over time, implanted electrodes also cause inflammatory reactions, or end up sheathed in brain cells and immobilized. They break down and, because the surgery is so invasive, they are difficult to replace.

In 1996, a pioneering Georgia Tech neurologist named Phil Kennedy won FDA approval to implant into the motor cortex of his human patients an innovative device that overcame some of these problems with movement. The device consisted of a pair of gold wires encased in a tiny glass cone. Filled with a proprietary blend of growth factors, the electrode induced nearby neurons to grow into it, which dramatically reduced the risk it would jar loose and cause scarring. The device could then be connected to electronics, capable of amplifying the neuron's signals and transmitting them out of the skull to a computer, where they could be analyzed.

Kennedy's first volunteer was a special education teacher and mother of two named Marjory, or "M.H.," who agreed to undergo the procedure at the very end of her life. Marjory had ALS, and could no longer speak or move, yet she demonstrated, remarkably, that she could turn a switch on and off just by thinking. But she was so sick that only seventy-six days later, she died. Next, in 1998, came Johnny Ray, a fifty-three-year-old Vietnam veteran and drywall contractor who awoke from a coma with his mind fully intact but unable to move anything except his eyelids. Remarkably, here, too, Ray demonstrated he could learn to move a cursor on a computer screen just by thinking and communicate by picking words or letters from a menu—though the process was slow and arduous.

As a third patient, Kennedy chose a young father and avid outdoorsman who'd been diagnosed with ALS more than a decade earlier and believed himself to be on the verge of becoming fully locked in: David Jayne.

If the doctor had been correct on that sunny Georgia day that David Jayne received his death sentence, by 1998 he would have long since been dead and buried. Even so, by the time David volunteered to undergo Kennedy's experimental procedure, ten years had passed and he was in the throes of a battle that he feared might finally destroy him.

David had watched the birth of his daughter, Hannah, cradled her gently in his large hands, still strong back then. He worked hard, attempting to earn money to secure his small family's future in the time he had left. And after confirming that his ALS was not genetic, David and his wife even had a second child—a son named Hunter. They wanted Hannah to have a brother.

In those early days, David tried not to think about what was coming. He operated on a "need to know" basis, focusing on the day at hand, and only what his neurologists told him was essential. Then one day, while he was descending an inclined parking lot, David's left leg collapsed beneath him without warning and he tumbled down a cement ramp. Soon such tumbles became frequent. He'd pick himself up, dust himself off, and reassure those around him it was nothing. But the monster had begun to spread. By 1993, David had lost the ability to breathe, swallow, and speak on his own. By then, however, advances in medication gave Jayne the option of outliving his death sentence. And new, more compact kinds of ventilators allowed him to survive outside the hospital. Jayne was in his early thirties, too young to give up and die.

Technology even existed to allow those unable to speak to communicate by using fingers or eye movements to select letters on a keyboard or screen, and translate them into audible speech using a speech synthesizer. David could still twitch three of his fingers, so there was hope. But the technology was new and expensive. And it would take his family two years to save up the $10,000 needed to buy a speaking device. In the process David was granted a horrifying glimpse of what might lie ahead for him were he to become fully locked in.

During that time, only his father and wife, Melissa, could come close to understanding the jumble of noises that emerged as he fought to twist his lips to push out words. Even they were often flummoxed. His children, toddlers at the time, would often burst into tears.

"Mommy, what did Daddy say?" they'd plead.

When David's voice synthesizer finally arrived in 1995, it changed everything. The package had been preloaded with several different voices, each one of which they assigned different personalities. At the family dinner table, he would entertain his kids by toggling back and forth between "Perfect Paul" and "Huge Harry," sometimes typing in with his working fingers the words "Clean your plate" and intoning them in the most ominous voice the computer could create.

But now, as the end of the millennium approached, David's body was failing him in new ways, and he worried that soon he would no longer be able to move his fingers to type.

"My ability to communicate and encourage others is the only thing that makes this godforsaken hell bearable," David would later write.

To find the ideal location for his electrodes, Kennedy placed

David in an fMRI machine to monitor his brain, and also fitted the muscles of his hands and arm with electrodes, a technique Hugh Herr and others in the field of biomechanics used to measure activations. Then Kennedy asked David to move his fingers and hands to the best of his remaining ability. By synching up the muscle activity with the location of neurons firing in David's motor cortex, Kennedy found his implantation target. He would implant electrodes into the hand area of the brain, so that even if David lost the motor neurons capable of conveying the signals out of his cortex and to his fingers, Kennedy could detect the signals in the brain, transmit them out of David's head, and convey them to the computer wirelessly.

Later, when David entered the hospital, a neurosurgeon sawed open his skull and inserted two glass-encased electrodes directly into his motor cortex, then closed the incision with dental cement and sent him home to recover.

But just a few weeks after the first surgery, David started getting crippling headaches. His incision had become infected. Antibiotics beat it back, but then infection returned. Eventually David underwent four more surgeries—not to install the electronics, but to remove the hardware and save his life. He required a skin graft to close the wound. By the time it was over, David recalls, "My head looked like a road map of incisions and all were infected."

In order to keep connected to the world, David would have to find another lifeline. He would have to find a way to survive until new, less invasive solutions came on line. It was a terrifying prospect.

Kennedy's research, meanwhile, continued. In 2004, Kennedy implanted another young Georgia resident, Erik Ramsey. Ramsey had suffered a catastrophic brain stem stroke in a car accident, which left him locked in at the age of sixteen. Ramsey was completely unable to move anything except his eyes, allowing him to indicate yes (up) and no (down). Inside, however, he remained very much aware and alive.

This time, Kennedy went a step further than he had attempted to go with David. During his presurgical fMRI scan, Kennedy was able to pinpoint consistent areas of neural activation in the vicinity of the motor cortex when Ramsey attempted to say phrases like "This is an elephant" and "This is a dog." Kennedy assumed these activation areas corresponded to the control centers that influence how the muscles of the lips, tongue, jaw, and larynx move to produce the sounds—movements that Ramsey could only imagine. Kennedy chose the areas of the densest signals as the bull's-eye where he would implant the electrodes.

This time things went smoothly. Collaborating with researchers from Boston University, Kennedy recorded distinct neuronal firing patterns from fifty-six neurons in Ramsey's brain, as he "spoke" different sounds. Thanks to the data collected from Ramsey, Kennedy and his collaborators continued to publish high-profile papers on the results in journals like *PLOS One* and *Frontiers in Neuroscience* as recently as 2009 and 2011. As Ramsey tried to "speak," the electrode picked up the impulses in his motor cortex and conveyed them to a computer that moved a cursor into different quadrants on the screen, varying the tone of the sound based on the cursor location. It was promising. But then Ramsey became too ill to keep participating in the research.

By then, the FDA had also withdrawn permission to use the devices in any more patients. Kennedy says the agency began asking him for more safety data, including on the neurotrophic factors he was using to induce neuronal growth. When Kennedy couldn't provide the data, the FDA refused to approve any more implants. The delay slowed Kennedy's efforts considerably, delaying any new implants until 2014.

It would thus be up to others to take it further.

Like most in the field, Gerwin Schalk followed Kennedy's pioneering efforts with interest. But he was well aware of the limitations of single-unit electrodes. As someone who had dealt in the absolutes of computer code—pages and pages of instructions simply spelled out in logical if/then statements—the limited reach of electrodes always bothered Schalk. Even the most technically advanced electrode arrays were capable of recording from only a few hundred neurons at the same time.

At a conference around the turn of the millennium, however, Schalk met Eric Leuthardt, a young neurosurgeon at Washington University, in St. Louis, who was using an obscure technique called "electrocorticography" (ECoG), or "intracranial EEG" (iEEG), that allowed the researcher to extract signals directly from the surface of the brain. Though the technique did require surgery to lift off the skull and access the outer layer of the brain, it did not require the surgeon to actually penetrate into the cortex itself and insert electrodes into the brain's gray matter; that made it far less risky. And since the electrodes were placed directly on the brain, one didn't have to worry about the skull refracting the signals and blurring the origin of the impulses. Certainly the resolution was not as precise and localized as you could get with single electrodes implanted directly in the brain, but after getting to know Leuthardt, Schalk believed he could get close if he collected data from enough spots across the top of the brain and used the magic of mathematics to triangulate the origins of different signals.

The best part was that Leuthardt had a seemingly inexhaustible supply of research volunteers: epilepsy patients like Cathy.

As Schalk and I stand by Cathy's bedside in Albany, she relates the series of harrowing experiences that led her to take extreme measures. A few years back, she almost drowned when a seizure hit while she was playing with Barbie dolls in the bathtub with her young daughter. Another seizure struck while Cathy was driving one day. She pulled over just in time to put on the parking brake, but when her muscles seized up, she slammed her foot down on the accelerator and her car hurtled down an embankment. The catalytic converter then exploded into flames.

"She's like a person with nine lives," her mother pipes in.

That is why Cathy's doctor, Anthony Ritaccio, has taken her off her antiseizure medication, removed the top of her skull, and installed the net of electrodes on her naked cortex. Once they locate the source of her seizures, they can determine whether they might excise that portion of the brain permanently without any undue side effects. But to locate the source of the seizures, Cathy has to have one. And so she sits in her hospital bed, waiting, with a web of electrodes on her brain and a bundle of wires coming of it. She is a captive audience, in other words. And she is bored.

This makes Cathy the perfect experimental subject for researchers interested in dreaming up random, interesting tasks and watching the human brain in action at a level of detail few ever get to see.

In recent years, Schalk, believe it or not, has taught teen epileptics, similarly bedridden and bored, to think their way through the byzantine, mazelike corridors of the video game Doom as they gun down monsters without ever touching a joystick. Schalk has taught patients to "type" specific letters on a keyboard, spelling out words, and send an e-mail without ever moving their hands or fingers.

Some researchers have gone even further. In 2012, a quadriplegic working with researchers at the University of Pittsburg used her thoughts, recorded by electrodes, to pick up a chocolate bar with a robotic arm, bring it to her mouth, and take a bite. Still, there remains much work to be done before these demonstration projects actually move to reality.

"You do see robotic arms being moved today," says Schalk's mentor, Jonathan Wolpaw. "But they're not anything that's anywhere near ready to be taken out of the lab. There's no BCI that you would now want to use to control a wheelchair on the edge of a cliff or to drive in heavy traffic. And until that happens, their use is going to be very limited."

Even so, as soon as he discovered ECoG, Schalk emerged as a vocal proselytizer for the technology. In November 2006, he attended a three-day conference on "Smart Prosthetics" in Irvine, California, and—much to the annoyance of many of those present—stood up in a session and began to expound on the wonders of ECOG. By claiming that he could pry information from the brain without drilling deep inside it—information that could allow a subject to move a computer cursor, play computer games, and even move a prosthetic limb—Schalk was taking on "a very strong existing dogma in the field: that the only way to know about how the brain works is by recording individual neurons," recalls Elmar Schmeisser, a military scientist who was sitting in the audience that day.

Many of those present, in fact, dismissed Schalk's findings as blasphemy and stood up to attack them. Only implanted electrodes of the type used by Hubel and Wiesel, Georgopoulos, Fetz, and Kennedy offered access to a signal robust enough to use.

But Schmeisser found Schalk's words electric. As a program officer in charge of pitching and overseeing projects at the U.S. Army Research Office, Schmeisser had come to the conference wondering how he might adapt BCI for use by his intended constituency. Not the handicapped—certainly not people with ALS—but able-bodied men and women in the military. Schmeisser was looking not to restore lost function, but to enhance it.

Now here stood Schalk detailing an obscure technique that he claimed offered access to a vast population of neurons in real time and far exceeded the tiny population of neurons scientists had traditionally studied using single-neuron recording—a technique that required, at the same time, far less drastic surgical measures.

Schmeisser, a tall retired colonel with a receding hairline, glasses, and a neck the diameter of a small tree, was a Renaissance man of sorts, whose interests ranged far and wide. Schmeisser had a Ph.D. in the physiology of vision and advanced belts in karate, judo, aikido, and Japanese sword fighting. But most important, he was a voracious reader of science fiction. One of his favorite authors was E. E. "Doc" Smith. In Smith's 1946 classic, *The Skylark of Space* (originally published in *Amazing Stories* in 1928), the author describes a futuristic helmet capable of detecting imagined speech and conveying it to someone else—Smith had called it "a thought helmet." Ever since the eighth grade, Schmeisser had dreamed it might one day be possible to create a device capable of reading one's thoughts—sentences imagined but not spoken—and then somehow transmitting them to another person.

When Schmeisser heard Schalk, "everything all of a sudden became possible," he recalls.

Schalk had not previously considered the possibility of decoding imagined speech. He had never heard of a thought helmet. But when Schmeisser approached him after the session and shared his ideas, the young Austrian scientist was intrigued. Schmeisser suggested he could sell the project and win funding, if it led toward his eventual goal of a completely noninvasive thought helmet.

It takes a special kind of courage to step in front of a room full of division chiefs, mathematicians, particle physicists, chemists, computer scientists, and Pentagon brass and ask them, with a straight face, to green-light a project aimed at promoting telepathic communication.

But not long after meeting Schalk, Schmeisser took the podium at the Army Research Office headquarters in Research Triangle Park, North Carolina, and set out to make this case. Wielding a metal pointer with theatrical flourishes worthy of his Japanese swordsmanship, Schmeisser flipped through PowerPoint slides and worked to persuade the thirty skeptical experts arrayed in front of him at a U-shaped table that technology had finally placed him within striking distance of his boyhood dream. Neuroscience pioneers were already connecting electrodes to the motor cortex of paralyzed patients, decoding their intentions and training them to operate prosthetics simply by thinking, Schmeisser told his audience. A number of new techniques would soon allow similar feats using far less invasive technology.

It was time, Schmeisser argued, to make the leap—to take this cutting-edge technology past the motor cortex and the sensing areas of the brain, and to decode the neural signals associated with something far more complex—ideas and imagined human language.

His listeners' reaction was not what he had hoped for.

"Could this really work?" Schmeisser remembers the committee asking him on that pivotal day back in 2006. "Show us the evidence that this could really work—that you are not just hallucinating it."

After conferring, the committee did agree to provide Schmeisser with $450,000—instructing him to return the following year if he could produce data to convince them his idea was more than a fantasy.

The first step, Schalk and Schmeisser agreed, was simply to demonstrate that the "imagined speech" could be detected at all—to prove, in the words of that skeptical army research review panel, that Schmeisser was "not just hallucinating" the possibility of a thought helmet.

What Schalk and Schmeisser were after was not just signals associated with the muscles used to produce speech—but signals associated with silent speech, the things we only imagine saying.

To look for it, Schalk and Leuthardt recruited twelve bored and bedridden epilepsy patients like Cathy to participate in their first set of experiments. The patients were presented with thirty-six words that had a relatively simple consonant-vowel-consonant structure, such as *bet, bat, beat,* and *boot.* Schalk and Leuthardt asked the patients to say the words out loud and then to simply imagine saying them.

They conveyed the instructions to the patients visually (written on a computer screen) with no audio, and again vocally with no video, to make sure that they could identify incoming sensory signals in the brain and remove them from pattern recognition programs. ECoG electrodes recorded a precise map of the resulting neural activity.

As one might expect, when the subjects vocalized a word, the data indicated activity in the areas of the motor cortex associated with the muscles that produce speech. The auditory cortex, an area associated with hearing, and an area in its vicinity long believed to be associated with speech processing, called "Wernicke's area," were also active at the exact same moments.

When the subjects were told simply to imagine the words, the motor cortex went silent. But here's the strange part: Remarkably, even when the subjects silently "imagined" the words, the auditory cortex and Wernicke's area remained active. Somehow simply imagining hearing or even just imagining speaking a word caused these areas of the brain to light up with activity. Although it was unclear why those areas were active, what they were doing, and what it meant, the raw results were an important start. There was something there when subjects imagined speaking. Something that could be detected.

Schmeisser presented Schalk's data to the army committee and asked it to fund a formal project to develop a real mind-reading helmet. As he conceived it, the helmet would function as a wearable interface between mind and machine. When activated, sensors inside would scan the thousands of brain waves oscillating in a soldier's head; a microprocessor would apply pattern recognition software to decode those waves and translate them into specific sentences or words, and a radio would transmit the message.

The words or sentences would reach a receiver that would then "speak" the words into a comrade's earpiece or be played from a speaker, perhaps at a distant command post. The possibilities were easy to imagine:

"Look out! Enemy on the right!"

"We need a medical evacuation now!"

This time the committee signed on. To maximize the chance of success, Schmeisser decided to split the army funding he procured between two university teams. The first team, directed by Schalk, was pursuing the more invasive ECoG approach, attaching electrodes beneath the skull. The second group, led by Mike D'Zmura, a cognitive scientist at the University of California, Irvine, planned to use ordinary electroencephalography (EEG), a noninvasive brain-scanning technique far better suited for an actual thought helmet.

It was a member of the second team, a renowned neuroscientist at New York University named David Poeppel, who made the next big conceptual advance. Initially, sitting in his office on the second floor of the NYU psychology building back in 2008, Poeppel now admits, he had no idea even where to begin. Poeppel had helped develop a detailed model of audible speech systems, parts of which were widely cited in textbooks. But there was nothing in that model to suggest how to measure something imagined.

For more than a hundred years, Poeppel reflected, speech experimentation had followed a simple plan: Ask a subject to listen to a specific word or phrase, measure the subject's response to that word (for instance, how long it takes him to repeat it aloud), and then demonstrate how that response is connected to activity in the brain. Trying to measure imagined speech was much more complicated; a random thought could throw off the whole experiment.

Solving this problem would call for a new experimental method, Poeppel realized. He and a postdoctoral student, Xing Tian, decided to take advantage of a powerful imaging technique called "magnetoencephalography" (MEG), to do their reconnaissance work. MEG can provide roughly the same level of spatial detail as ECoG but without the need to remove part of a subject's skull, and it is far more accurate than EEG. The downside, of course, is that you can't move much, or it messes up the signal.

Poeppel and Tian guided their subjects into a three-ton, beige-paneled room constructed of a special alloy and copper to shield against passing electromagnetic fields. At the center of the room sat a one-ton, six-foot-tall machine, resembling a huge beauty parlor hair dryer, that contained scanners capable of recording the minute magnetic fields produced by the firing of neurons inside the head of anyone sitting beneath it. After guiding subjects into the device, the researchers would ask them to imagine they were speaking words like *athlete, musician,* and *lunch*. Next, subjects were asked to imagine hearing the words.

When Poeppel sat down to analyze the results, he noticed something unusual. As a subject imagined hearing words, the auditory cortex lit up the screen in a characteristic pattern of reds and greens. That part was no surprise; previous studies had linked the auditory cortex to imagined sounds. However, when a subject was asked to imagine speaking a word rather than hearing it, the auditory cortex flashed an almost identical red and green pattern. It was similar to the trace that Schalk had spotted during the initial phase of the experiment. But Poeppel's results were even clearer—the "something" Schalk and Leuthardt had spotted using ECoG bore a striking similarity to what you would see in real speech. Was it possible that the little voice we sometimes "hear" in our heads might in fact actually be a little voice we hear in our heads?

Scientists had long been aware of an error-correction mechanism in the brain associated with motor commands. When the brain sends a command to the motor cortex to, for instance, reach out and grab a cup of water, it also creates an internal impression, known as an "efference copy," of what the resulting movement will look and feel like. That way, the brain can check the muscle output against the intended action and make any necessary corrections.

Poeppel believed he was looking at an efference copy of speech in the higher-level processing areas of the auditory cortex. "When you plan to speak, you activate the hearing part of your brain before you say the word," he explains. "Your brain is predicting what it will sound like."

The potential significance of this finding was not lost on Poeppel. If the brain held on to a copy of what an imagined thought would sound like if vocalized, it might be possible to capture that neurological record and translate it into intelligible words. But building a thought helmet would require not only identifying that efference copy but also finding a way to isolate it from a mass of brain waves. It was there that Schalk's expertise would come into play.

In the end, decoding brain signals from millions of neurons has proven largely a numbers game. Since the program began, Schalk and his collaborators have demonstrated they can tell the difference between vowels and consonants. But the detection has never been perfect. After years of efforts and refinements to his algorithms, Schalk's program was able to guess correctly 45 percent of the time. (Chance is 25 percent.) But rather than attempt to push those numbers up toward 100 percent, Schalk has focused on continuing to build on the complexity of what he can decode.

From differentiating between individual vowels and consonants, he has moved on to differentiating between vowels and consonants embedded in words. Then to individual phonemes—in other words, any sound that distinguishes one word from another, such as the *m*, *c*, and *t* sounds in *mad*, *mat*, *cad*, and *cat*.

That is about as far as Schalk had gotten when Schmeisser's funding ran out. The end results were a long way from Schmeisser's thought helmet. But actually building the device, Schalk argues, was never the point.

"In 2006, all of this was completely speculative and completely science fiction," Schalk says. "Clearly that is not the case anymore."

And indeed, for Schalk some of the most exciting developments have occurred since the project ended.

From Cathy's bedside, I follow Schalk down a hallway, into an elevator, and out to a parking lot. Then Schalk and I make a quick drive down the street to a boxy, concrete building. In his sparsely furnished office, Schalk sits me down in front of a large computer screen, next to a pair of speakers.

Schalk pulls up a mass of brain signals, squiggly lines and different kinds of charts, on the screen in front of me. Then he flips on the speakers. Over the course of many months, Schalk explains, he carried speakers into hospital rooms and played the same segment of a Pink Floyd song for about a dozen brain surgery patients like Cathy. As the patients listened to the song segment, Schalk recorded the activity of neurons in their auditory processing areas of the brain. Then Schalk handed the file of recorded brain activity over to the UC Berkeley lab of Robert Knight for processing.

Would I care to hear what Knight and his team were able to pull out of that mass of brain data? Schalk presses a button.

It's uncanny. A bass begins to thump urgently from the speakers like the furious beating of a human heart. It's slightly muffled, as though heard from underwater, but it's clearly a bass. A plaintive guitar echoes through an effects pedal, its notes accelerating with each new phrase. The pacing is frantic, but there's a hypnotic quality to the music coming out of the speakers. I recognize the song immediately—it is the mesmerizing and haunting tones from Pink Floyd's "Another Brick in the Wall, Pt. 1" (chosen, according to Schalk, because something that felt so psychedelic seemed appropriate). Aside from the vague muffling, the song is identical to the song I used to listen to in high school. But this version comes from brain waves, not music. I feel chills run up and down my spine.

"Is it perfect?" Schalk asks. "No. But we not only know that he's hearing music, not only what genre or whatever, you know the song—I mean, clearly you know the song. So, this is pretty crazy. It used to be science fiction, but it's not anymore."

This feat is possible thanks to the same discovery that allowed bioengineers to hook prosthetics up to the motor cortex: the discovery that most neurons have a "tuning curve." Just as Georgopoulos discovered that some neurons in the motor cortex have a direction of movement they are most likely to respond to, researchers specializing in hearing have since demonstrated that neurons in the auditory cortex also respond preferentially to different kinds of sounds. Instead of direction and velocity, these neurons fire more robustly in response to specific tones and amplitudes. Hit an individual neuron's sweet spot in the auditory cortex by playing the right tone, Knight and his team knew, and it fires robustly, just as that neuron in the visual cortex of Hubel and Wiesel's cat fired robustly in response to a line at a specific angle. Move away from a neuron's preferred tone (determined by some combination of genetics and experience) and the neuron's firing rate will slow, just as a neuron in the motor cortex slows as an intended direction of movement diverges from its preferred angle.

Schalk and his Berkeley collaborators are now attempting to discern whether patients are imagining reciting the Gettysburg Address, JFK's Inaugural Address, or the nursery rhyme "Humpty Dumpty" just by looking at brain data—and attempting to reproduce it artificially using the same techniques.

There is another promising development. In 2014, Phil Kennedy jumped back in the race to decode imagined speech. Still unable to implant into new patients in the United States, Kennedy feared his twenty-nine years of research would "die on the vine" if he didn't do something drastic. So Kennedy traveled to Belize and paid a neurosurgeon $25,000 to implant electrodes into his own motor cortex speech area. Since he could still talk, and was conducting the experiment himself, Kennedy figured, he would know exactly at what point he was imagining the speech, allowing him to home in on the brain signals firing at precisely the moment he imagined them.

After returning home to Duluth, Georgia, Kennedy began to toil largely alone in his speech lab, recording his neurons as he repeated twenty-nine phonemes (such as the vowels *e, eh, a, o, u,* and consonants like *ch* and *j*) out loud, and then silently imagined saying them. He did the same with about 290 short words such as *dale* and *plum.* He also spoke phrases that had different, especially distinct, combinations of phonemes: "Hello, world," "Which private firm," and "The joy of a jog makes a boy say wow."

Kennedy had hoped to live with the implants in his brain for years, collecting data, improving his control, and publishing papers. But the incision in his skull never closed entirely, creating a dangerous situation. After a few weeks of collecting data, in January 2015 Kennedy was forced to ask doctors at a local Georgia hospital to remove the implants, which would allow the incisions to close. The bill came to $94,000. Kennedy submitted the claim to his insurance company. (He says it paid $15,000.) Still, he managed to get enough data to resume his research and hopes to yield insights that will complement Schalk's parallel effort in the auditory cortex.

"I got away with it, so I'm happy," he says. "I had a few bumps and bruises after the surgery, but I did get four weeks of good data. I will be working on these data for a long time."

Almost thirty years after receiving his diagnosis, meanwhile, David Jayne is still with us. When he lost the ability to move his fingers, he began using an infrared switch mounted on glasses and was able to raise his eyebrows to activate it. Later he volunteered to serve as a guinea pig for another Phil Kennedy project, which led to the development of the device he currently relies upon, a Bluetooth electromyogram (EMG) sensor on his jaw that detects electricity generated by muscle contractions and relays it to software that interprets it as either a keystroke or a mouth click. Back in 2000, the device consisted of two pieces of equipment, together about the size of a loaf of bread, with an additional laptop and ten yards of cables and wires. Today it is wireless and compact.

It's enough, it seems, for now. Jayne is currently capable of typing nine words a minute. In recent months, Jayne says, he has discovered a new diet and a new rehabilitation protocol, which he claims has begun to restore some limited movement. He can now smile, move his neck, and flex the muscles in his legs. He posts his progress regularly on Facebook, where he has almost 3,300 friends.

Over e-mail, I asked Jayne what he thought about Gerwin Schalk's efforts and the continued experiments of Phil Kennedy.

"I'm in awe of Phil," he wrote. "Talk about walking the walk."

Of a speech prosthesis, he had this to say: "If proven I would embrace it absolutely. I'm assuming this technology will be nearly spontaneous? Learning how to keep my imagined mouth shut would be extraordinarily difficult. 'Good Lord, that woman has a great ass! Did I say that out loud???'"

It's undeniable David has continued to *live* his life. He has been married, if you can believe it, three times. (He has retained sensation and touch, so he still has sexual function.) He has fought and reconciled with numerous members of his family. His sister Sue Ann says that his ability to communicate is now so advanced that they have even been able to fight and then reconcile, and of late they have had many deep conversations.

"Oh, let me tell you, it's the full rich human experience, once he was able to communicate," she says. "Many times I thought about unplugging it [the communication device]. Just shut up!"

The fact that one of David's loved ones is able to respond with a comment so flippant is a more powerful testament to the effectiveness of the technology than anything else I encountered.

But Jayne doesn't want anyone to misunderstand.

"Of course I had tons more fun pre ALS," he writes.

His mother is more frank.

"Anything you write about this condition is good," she told me. "But please don't make it seem like it's a rose garden. It can do terrible things to families and not a lot survive the changes. David has had some rough times; I don't know if you know how rough a time he has had."

Jayne insists he has found some measure of contentment.

"Nothing in my life is the same except for my determination and love of new experiences," he writes. "This is my life. I have accepted that. So I am at peace and that generates happiness."

Even so, David, for one, maintains that his ability to communicate is just as important to his survival as his ability to breath.

"I don't scare easily, but I have orders to pull the plug if I ever lose the ability to connect with the world," he writes. "I am not interested in discovering the quality of life that would be. As for life now, what I do generates uncomfortable lags, but the high of knowing you encouraged someone is better than any pharmaceutical I have done."

PART III

THINKING

CHAPTER 7

THE BOY WHO REMEMBERS EVERYTHING

VIAGRA FOR THE BRAIN

Someone was bound to get hurt. But if you knew the cast of characters in the midwestern neighborhood where Tim Tully and his five rambunctious Irish Catholic brothers and sisters grew up, you'd understand.

Washington, Illinois, was the kind of place where dozens of kids ran through streets lined with neatly trimmed hedges and modest split-level houses in the hot summer months spraying each other with water guns, wrestling, and playing manhunt. Like most packs, there was often an alpha male or female, and nothing helped you climb the hierarchy for the day like a good showing in the local neighborhood's version of "chicken." In the summer they used their bicycles, with two competitors powering toward one another head-on at full speed. The goal was to run your opponent off the road. In the winter, Tully and his pals put their bikes away and got out their sleds.

That explains the events of that snowy Christmas Day in 1968. Presents had been opened and now it was game time. So the gang gathered at the top of a hill behind the Tully house to play a variation of chicken they'd invented more suited for this unique terrain. The object was to drive other sledders over a fearsome overhang the kids called the "cliff." The playing field consisted of a gentle slope running north to south along three yards. (The Tully house was in the middle.) At the back of these yards was a ridgeline, below which the terrain dropped twenty-five feet, at an angle of roughly 80 degrees. This fall concluded about a third of the way down, with a sheer, ten-foot drop onto the frozen creek below. Most sledders, when forced off the "cliff," deployed a standard survival tactic. The best approach was to roll off your sled and try to grab on to a tree to slow your descent. It wasn't so bad. After retrieving your rig, you made the humiliating climb back to the top, pulling yourself up the steep hill and steadying your ascent by using roots and branches as handholds. At the top, after you endured a brief bit of razzing, you had a chance for revenge.

That day, Tully, fourteen at the time, was having a blast, barreling down the hill on his Flexible Flyer, yelling and raising hell—when a good blow from a competitor sent him hurtling over the side of that steep cliff headfirst. Feeling daring, Tully made a snap decision to forgo normal evacuation procedures. Instead, he decided to enjoy it. There was, in his estimation, between two and three feet of snow on the ground in front of him, which he figured would cushion the reckoning when he finished his descent at the bottom. Clutching the sides of his sled tightly, he hunkered down. It probably would have worked were it not for a buried tree stump, obscured by the snow. When the sled hit the stump, Tully was catapulted forward into a series of airborne acrobatics and ground-smacking somersaults so improbable they would become a permanent part of neighborhood lore.

In the process, Tully slammed his knee into his face and knocked himself unconscious. Months later, after the snow had melted, he would return to the site and find one of his front teeth, still attached to its root, lying in the grass. Tully had no memory of losing it. In fact, Tully had no memory of the events of that day at all. Remarkably, when he'd awoken groggy and concussed in a bed back home, his mouth swollen shut, Tully was certain it was the Friday two weeks prior. Somehow, he had "lost" the previous fourteen days, as surely as the Henry Molaisons of the world lose every single new hour of their lives. It was as if the accident had wiped part of Tully's hard drive clean. Tully's memory gap was so solid, his insistence on the incorrect date so unyielding, that his parents rewrapped his Christmas gifts so he could experience the joy of opening them anew.

In the years that followed and into his college years, Tully would often contemplate the mysteries of recall and wonder about that sled ride. How was it that a knock on the head could cause his memory of *two full weeks* to disappear? And if new memories were discrete things that could be separated from the rest of our experiences and somehow erased just like that, might it be possible to do the opposite—to capture new experiences and instead somehow imbue them with the instant clarity and permanence of a painting or a photograph? Might it be possible to reverse the inevitable decline in recall we all seem to face as we age? Was there a way to untangle the mysteries of memory?

Tim Tully and I are standing in the middle of an expansive, sun-draped atrium, surrounded by glass windows rising several levels. It's hard to imagine the man next to me as the young teenager with the missing front tooth from tiny Washington, Illinois.

Both Tully's neatly cropped beard and the thick mop of graying hair he's swept rakishly across his forehead lend him an air of respectability. And inside this sprawling 200,000-square-foot former Nokia cell phone plant perched on a hill on the outskirts of San Diego, Tully doesn't have to brave head-on collisions to win the rank of top dog. He runs the show every day.

As we gaze up at the tiers of glass-windowed laboratories, I can see young, white-jacketed researchers buzzing about, performing protein crystallography and other feats of scientific magic. Nearby, precision robotic arms with long metal pincers grip tiny plates, each containing one of 1,536 different pharmaceutical compounds. The arms swivel the trays back and forth, moving them between incubators and imagers.

In another area, biochemists are constructing an 800,000-molecule-strong chemical library, much of it from scratch, which can then be used to create new drugs.

"We keep our nose down; we don't need to raise money," Tully tells me. "But this is the real deal."

This is where the journey that began on that steep cliff more than four decades ago has led. Today Tully is at the forefront of efforts not just to unlock the many secrets of memory, but to go far beyond. With the backing of a reclusive billionaire who is prepared to kick in up to $2 billion to fund the scheme, Tully is attempting to create and win approval for drugs that might someday allow all of us to function with superior recall and minimize those dreaded "senior moments." He is, in other words, attempting to artificially produce the opposite effect of a tooth-rattling, concussion-inducing knee to the face. Tim Tully is making a "memorization pill."

That it is possible to find a way to give us all something like photographic memories, Tully has no doubt. He has already done so in fruit flies, mice, and other mammals.

"Oh, we'll find it," he says. "Don't worry about that."

But don't take his word for it. The evidence to prove this, Tully argues, isn't hard to locate. All you have to do is look to history.

Most of us struggle with memory at times. We scrunch up our foreheads and strain to call up details we should know. Where did we put the car keys this morning? What was the name of that woman we met at the cocktail party? What *did* happen over the past two weeks?

In the worst-case scenario, we lose our memory altogether. Dementia and Alzheimer's are lonely good-byes, often slow and steady enough to allow those condemned to suffer them the cruel knowledge that everything is gradually slipping away.

I watched from afar as this happened to my grandfather Mannie, a brilliant silver-haired physicist who smoked a pipe and sat with me watching my beloved Boston Red Sox play ball on the television. Mannie had immigrated from Eastern Europe, grew up poverty stricken in Harlem, and then rose to become the first Jewish top science advisor to a U.S. president. In his final years, he slowly disappeared into a confused mental fog; eventually my proud and committed grandmother Nora had to admit she needed help. Toward the end, Mannie would wake up early in the morning in a frenzy, insisting to the Haitian health-care aides on duty at his long-term residence in Brooklyn that he was late for a meeting at the White House.

I used to worry about my father, who I have watched daily search in vain for his house keys since I was a child. But lately I've begun to notice the limits of my own memory—were it not for my wife, we would never leave the house. With apprehension sometimes, I wonder what lies ahead.

Some suggest dementia was the first deadly disease described in detail in Western literature. In *The Odyssey,* Homer tells the story of Odysseus's father, King Laertes. When Odysseus returns home after a twenty-year absence, he finds his father wearing dirty old clothes, laboring on a farm, entirely oblivious to the threats to his estate.

But just as memory loss is a staple of history, so too, it turns out, is the opposite. Every once in a while, someone is born whose brain seems to operate by a different set of principles than the rest of ours do—someone whose memory defies normal limits, inspires the Tim Tullys of the world, and suggests that, just maybe, it doesn't have to be this way.

In the nineteenth century, there was "Blind Tom" Wiggins, an enslaved African American savant who was said to be able to repeat conversations up to ten minutes in length, learned to play and compose on the piano, and then became world famous for his ability to reproduce from memory any piece of music played for him just once. He toured the world and performed for President James Buchanan at the White House. More recently, there was the American Kim Peek. He could read a book in an hour and recall its entire contents. Before he died in 2009, Peek had read as many as twelve thousand books, remembered most of them, and inspired a movie (*Rain Man*).

In the twentieth century, the world learned of a Russian newspaper reporter named Solomon Veniaminovich Shereshevsky, who could recite decades-old conversations and speeches word for word, and was able to memorize long numbers, mathematical formulas, and portions of books in mere minutes.

"I simply had to admit," wrote the Russian neuropsychologist Alexander Luria of his famous subject, "that the capacity of his memory had no distinct limits."

Today we have Jake Hausler. When he was two, Jake began pointing at cars during walks through his suburban New Jersey neighborhood with his mother, Sari, and reciting seemingly random numbers. Then one day, Sari realized the pattern: Jake was referring to the numbers on the inspection stickers placed in the corner of each car's front windshield. He had memorized them. Soon after, Sari's husband, Eric, realized Jake knew the aisle and exact location of every product in the local ShopRite.

By five, Jake could tell you not only the capitals of every state, but each state's flag and most of their state birds. From his booster seat on the way home from school one day, he reeled off the day of the week, and the weather for every day going back months, as well as which member of his kindergarten class had been called up to the front of the room that day to note it on the board. He then demonstrated to his astonished mother that he could deliver this information either chronologically, or by listing off the names of his classmates in alphabetical order.

People like Blind Tom Wiggins, Kim Peek, Solomon Shereshevsky, and Jake Hausler hint at untapped mental mechanisms waiting to be unlocked in all of us. It's only in recent years, however, that scientists like Tully have truly begun to unravel these mysteries at a resolution that might allow us to actually tap into this potential, and find a way to unlock it in all of us. It's only now that we have the brain-scanning tools, the genetic analysis, and other diagnostic mechanisms that give someone like Tully a real shot at cracking the code.

Most trace the birth of modern memory science to the late nineteenth century, when a thirty-something German philosopher with a thick, dark beard and an uncommon (some might say masochistic) determination traveled alone to Paris.

The philosopher was Hermann Ebbinghaus. Upon his arrival, he rented a room overlooking the roofs of the city and set to work on a task so tedious, thorough, and eventually revolutionary that his resolve would elicit superlatives from generations of future scientists. (William James called his actions "heroic.")

Ebbinghaus was determined to remain until he had precisely quantified the far limits of his memory—whatever it took. Doing so required extreme measures. To ensure that previous associations would not cloud his experimental results, Ebbinghaus created 2,300 new combinations of vowels and consonants he called "nonsense" words. He wrote each one on a separate slip of paper, and then drew pieces of paper by chance to construct lists of nonsense words from seven to thirty-six items long. Then he set to work memorizing each list.

Ebbinghaus was interested in understanding and characterizing the power of the most basic tool we use to remember things—repetition. He stuck to a rigorous procedure on each trial, so as not to corrupt his conclusions: First to a metronome and then to the ticking of his watch, Ebbinghaus read off each word on the list at a rate of exactly 150 words a minute. He repeated each full list until he could recite it from memory. He timed the intervals between learning sessions. He made sure to do the work at the same time every day. All the while, Ebbinghaus meticulously plotted both the accuracy of his new memories as they formed, the amount of repetitions needed to form them, and what he called "the curve of forgetting," detailing how and when the memories began to fade. By doing so, Ebbinghaus produced among the first concrete rules characterizing exactly how repetition can be used to form a new memory.

One of the first things Ebbinghaus demonstrated was that the time required to memorize nonsense syllables increases sharply as the list grows longer. Lists about six or seven items long could often be learned in only one presentation, while longer lists required more practice (a phenomenon due in part to the limited capacities of "working memory"). Ebbinghaus also demonstrated that we are more likely to remember items that appear at the beginning or end of a list than in the middle (which he called, respectively, the "primacy" and "recency" effects). He also showed that even practice far below that needed to memorize a word can still make it easier to learn that word later on. Memory is cumulative.

One of Ebbinghaus's most important and fundamental observations, however, was that while repetition itself can be sufficient to make memories stick, the *kind* of repetition and the *spacing* of that repetition make a big difference. Ebbinghaus found that the most effective way to memorize the material was to distribute shorter learning sessions over time, rather than cramming them into a single continuous session. As any college student knows, cramming for a test at the last minute might improve your score in a pinch, but it is also a good way to quickly forget everything you just learned soon after the test is over. It's far better to space out learning sessions, and repeatedly expose oneself to something we'd like to recall after intervals of rest.

Why does the timing of this repetition matter? The German psychologists Georg Müller (a former student of Ebbinghaus) and Alfons Pilzecker suggested the answer a few years later. After replicating Ebbinghaus's results, they showed that they could actively impair the formation of a new long-term memory if they exposed a subject to a new nonsense word too soon after the initial memorization took place. This worked, they hypothesized, because the brain needed time to perform the mysterious alchemy needed to transform new memory traces into permanently stored long-term memories, a process they called "memory consolidation." When they exposed their subjects to new nonsense words too soon after presenting the initial word, the scientists disrupted this process of consolidation. The brain lost its tenuous grasp on the first word the subject had just learned. The memory slipped away before it had solidified. But if Müller and Pilzecker instead repeated that first word, it actually strengthened consolidation and improved memory.

Even at the time, other scientists noted that the "consolidation hypothesis" provided a possible explanation for a mysterious phenomenon of memory loss often observed in patients who sustained head trauma—the very same mystery Tully would later wonder about after that sledding accident erased two weeks of recall on that snowy Christmas Day in 1968. Was it possible, these scientists wondered, that a blow to the head might somehow interfere with the brain's ability to transmute the fleeting impressions of the recent past into memories capable of lasting a lifetime?

It was a theory that was soon forgotten in the world of science, until almost five decades later.

In 1930s, two Italians devised a method of using electrodes to deliver electric shocks to the heads of mentally ill patients. The procedure, called "electroconvulsive therapy" (ECT), caused seizures that, for reasons that remained elusive, helped relieve symptoms associated with schizophrenia, depression, and other psychiatric disorders. ECT could be hard to watch—it caused violent, writhing convulsions. But ECT also had a bizarre side effect that, in the world of public health, helped overcome any moral objections: It almost invariably caused memory loss, most powerfully for events occurring just before or after the treatment. That meant the patients couldn't remember the procedure.

By the late 1940s and early 1950s, ECT was widely used around the globe. Memory researchers recognized that ECT was the perfect tool to test the consolidation hypothesis. One set of researchers ran rats through a series of different compartments and then zapped their brains with electricity. Remarkably, they discovered that the rats that got a shock seconds after running through the compartments seemed to lose all memory of the event. Yet the rats shocked after a sufficient interval of time seemed to retain the memory. The rat brains indeed just needed time to consolidate the memory of that experience into long-term storage. Given the time, it was far harder to dislodge a consolidated memory and make it disappear.

Still, some key pieces of the puzzle remained missing. After all, who would deny that we can recall some memories in vivid detail after experiencing the thing remembered only once? Why is it that a father can vividly recall the birth of a child? Or that most of us remember exactly where we were when the Twin Towers fell on September 11, 2001? Why isn't repetition needed in these cases?

One day in the late 1950s, a twenty-four-year-old graduate student named James McGaugh was conducting research at the University of California, Berkeley, library when he stumbled across an obscure yet fascinating paper from 1917.

Inexplicably, when researchers injected rats with the stimulant strychnine before placing them in a maze, it seemed to significantly improve their ability to form new memories. The second time through the maze, these rats went down fewer dead ends and finished faster than their peers.

McGaugh knew that the cause could have been any one of a million things—perhaps the drug improved visual acuity, attention, or motivation. (Strychnine is, after all, a stimulant as well as a poison.) But McGaugh had been closely following the experiments with ECT and the exciting ideas emerging about memory consolidation. Was it possible, he wondered, that this drug somehow had the opposite effect—that the stimulant enhanced memory consolidation? It was a long shot, but McGaugh realized there was a straightforward way to test for it that would eliminate these confounding factors. McGaugh could repeat the experiment he'd read about. But instead of giving the rats the drug before the trial, he could wait until *after* they had run the maze.

The approach made sense. After all, scientists had demonstrated you could block consolidation by administering an electric shock after the rats had run the maze. But McGaugh's advisor didn't see it that way. He practically laughed the young researcher out of his office.

"You want to give the drug afterward?" his advisor asked incredulously. "Afterward, the learning is over," the advisor scoffed. "What a crazy idea!"

Lucky for McGaugh, his advisor was at that very time packing for a sabbatical halfway around the world. When the advisor left the country a couple of weeks later, the young graduate student went ahead with his plans. He dosed half his rats with a small amount of strychnine after they had run the maze, and gave the other half a dose of saline solution.

"I get chills whenever I think about it, honestly," McGaugh says today of what happened next.

The impact was hard to miss. The rats that had been given the strychnine immediately after running the maze remembered the maze the second time through far better than the regular rats. They went down fewer dead ends and needed fewer subsequent training trials to learn the whole route through. Somehow the stimulant caused them to remember it better.

"My God, look at this," he recalls thinking. "The damn thing worked!

"I can remember feeling that I was walking about four feet off the floor because nobody had ever seen that before."

It's now known that a wide array of stimulants can modestly enhance memory consolidation. And McGaugh has spent the last fifty years examining why.

McGaugh reasoned that there must be something in the body that was already designed to do what the stimulants were doing—that giving these rats stimulants was somehow hacking into a biological system honed over millennia by the process of survival of the fittest. But why would an organism's chance of surviving to pass on its genes increase if it were born with a brain capable of influencing the strength of a memory *after* the event occurs? Why not simply make *all memories* strong—instantly? What survival advantage could selective, after-the-fact memory consolidation possibly confer?

When McGaugh examined his own memories, he realized that there were certain experiences that stood out. Some were happy—like that eureka moment in the laboratory. Others were sad—like the somber drive he made back to Eugene, Oregon, from Portland the day he learned JFK had been shot. McGaugh realized in a flash what his most vivid memories had in common: Many of them had deep emotional significance. We remember experiences that are meaningful. That got McGaugh and his research team thinking.

As scientists, they knew that the body has a stock response to emotion: It releases stress hormones, regardless of whether the emotion is positive or negative. Could it be that the stress hormones were the modulator on memory that he was seeking?

To find out, McGaugh's team took rats in the lab and applied an unpleasant stimulus that the rats would be motivated to avoid if they remembered it. They applied a mild electric shock to the rats' feet in a certain part of their training alley. After they administered the shock, the team administered a shot of adrenaline, a hormone that is released from the adrenal glands and is best known for producing the burst of energy associated with the fight-or-flight response. When they administered the adrenaline two hours after the shock, there was no memory enhancement. But if they delivered a shot right after the event, there was a profound enhancement of memory. There was no mistaking it—adrenaline augmented the storage of memory. Later McGaugh would follow the trail and demonstrate that a wide array of neurological processes and hormones involved in stress could enhance memory consolidation.

These hormones, McGaugh says, function "like a volume switch."

"There's a long-term understanding that forgetting some things is good," McGaugh says. "Otherwise we would have too much going on at one time. You don't want to remember what your foot feels like at this exact moment, for instance. But our ancestors needed to know where the food was, they needed to remember where the predators were to survive. They needed a way to remember emotionally significant events."

Over the years, McGaugh's findings have been of deep interest to some pharmaceutical companies attempting to develop memory-enhancing drugs. So far, many of these efforts have been slowed by unwanted side effects. (Some stimulants, for one, are addictive.) But McGaugh's work in this area would also inspire a generation of new scientists intent on drilling down even deeper to gain insights into what may actually be happening on the molecular level. Perhaps there was a more direct way to control this volume switch.

That brings us back to Tim Tully.

Dart NeuroScience, where Tully serves as executive vice president of research and development and chief science officer, is perched on a high ridge overlooking miles of green, rolling hills and the distant stripe of Southern California's Interstate 15 far below, a few miles east of San Diego's golden Pacific beaches.

It's one of those idyllic, Southern California afternoons on a balmy, cloudless December day when I visit, a welcome respite from the miserable East Coast winter I've escaped back home. In front of the main entrance an oversize company flag billows in a supple breeze, with its blue crescent and a cluster of small dots meant to represent molecules. Tully meets me in the lobby, dressed casually in a pair of black jeans and a button-down shirt, rolled up at the sleeves.

He shows me the outdoor Zen garden with its burbling fountain, a place where a scientist might retreat for a little solitude to contemplate vexing questions (if he or she has the mental strength to block out the drone of the freeway drifting up from far below). We peer into the fitness room with its immaculate equipment, straight out of the box.

Tully leads me into his company's custom-built hundred-person auditorium, where state-of-the-art video facilities stream live lectures for the staff from the brightest minds at the nearby University of California, San Diego, and the Scripps Institute (two lectures a week in neuroscience, one in cognitive sciences). His pride in the still relatively new results of a multimillion-dollar gut renovation seems to extend almost down to the level of houseplants.

"It's an incredible opportunity," he tells me. "I mean look at it."

All of this is possible thanks to the vision and ambition of a secretive billionaire named Ken Dart, who read an article about Tully in *Forbes* magazine and speculated on the impending arrival of what it dubbed "Viagra for the Brain." At the time, Tully held a secure, quiet job and his own lab at Cold Spring Harbor Laboratory, a well-regarded not-for-profit research facility in Long Island, New York, run by the Nobel laureate James Watson. There Tully attempted to unravel the biochemical and genetic basis of memory. He published often, engaged in the sharp-elbowed battle for prestige and scientific credit, and delivered lectures at academic conferences on topics most would consider obscure, like "CREB-binding" proteins and "the expression patterns of genes."

But on the side, Tully had founded a small company with Watson, the legendary co-discoverer with Francis Crick of DNA's double helix structure. The goal of Helicon Therapeutics was no less ambitious than that of Tully's current company. But the budget could not match the dream of a memory pill and progress was slow. Like a spate of other similar companies launched around the turn of the millennium, Helicon was in and out of financial trouble, and never did seem to get much traction. But Dart, heir to the Styrofoam empire, had been intrigued enough by what he'd read in *Forbes* to make a modest initial investment in Helicon. And in 2007, the straight-talking magnate approached Tully with a more ambitious proposal: How would he feel if Dart started a new company, bankrolled the entire effort himself, and put Tully in charge? Dart seemed to be offering Tully access to virtually unlimited resources in pursuit of his goal to create the world's first "memory pill." But there was a condition: Tully would have to agree to work full-time for Dart.

Ken Dart is not particularly known for profligate spending. In fact, to hear some members of the U.S. Senate tell it, he's a modern-day Ebenezer Scrooge. A prominent "vulture capitalist," known for feeding on distressed assets, Dart has been in the news in recent years for moving to the Cayman Islands to avoid paying taxes, for refusing to compromise with the near-bankrupt nations of Argentina and Greece when he was a creditor holding their debt instruments, and for bankrolling research into cryonics. News reports suggest Dart funded that effort in the hopes of remaining technically "alive" forever so he can escape someday paying estate taxes.

Tully was blunt. "You have to realize it's going to cost you one hundred million dollars a year for twenty years before you make a profit," he warned Dart. In other words, about $2 billion.

"That's about what I was expecting," Dart replied.

Dart had passed Tully's "flinch test." So Tully took the job and set to work building his fantasy facility.

Tully is now standing outside his company's new deli/lunchroom, which, he informs me, can be opened up to seat three hundred. I may have seen it on the news—it's used for more than just meals. Once a year the lunchroom serves as the arena for the Dart NeuroScience "Extreme Memory Tournament (XMT)," an annual contest that draws some of the world's top mental athletes from far and wide to this out-of-the-way San Diego office park to show just how far memory can be pushed. The 2015 contest offered $76,000 in prize money and drew finalists from Germany, Sweden, the Philippines—even from the up-and-coming Mongolian national team. The feats of recall on display were truly astounding. To win, contestants tried to remember 80 random digits, 50 randomly ordered words, 30 pictures, and 30 face-name pairs, among other things.

To win the two-day, 2015 tournament finals, Johannes Mallow, a diminutive, thirty-three-year-old German, memorized 80 digits in 21.01 seconds. Fellow German Simon Reinhard set the record for cards—memorizing 52 cards in 23.34 seconds. Enkhjin Tumur, a seventeen-year-old from Mongolia, memorized the order of 30 pictures after studying them for just 14.40 seconds. That's about the time it probably took you to read the last couple of sentences.

Though the XMT has garnered Dart headlines in the *New York*

Times, Guardian, and presumably across Mongolia, it is far more than just a branding tool. Each year, Dart and a team of academics from Washington University confer with the contestants and gather data as part of a larger search for the elusive white whale of memory science: a complete and total natural—a modern-day Solomon Shereshevsky, Blind Tom, or Jake Hausler. Someone, in other words, who has inherited the genes for a brilliant memory. Someone who can perform better than James McGaugh's strychnine- and adrenaline-dosed rats performed—but without the drugs.

The memory tournament is just one part of an unprecedented global hunt for genetic variants that offer superior memory abilities. Today more than ever, Tully is convinced that by understanding the genetics of those with superior memory, he can find ways to enhance it in the rest of us. If Tully finds mutations in genes coding for specific proteins involved in memory, he believes, he can replicate the effects of the mutation with small molecular compounds created using his 800,000-strong molecular library.

Still, even before Tully affixed Dart's name to the Extreme Memory Tournament, he knew that most of the contestants likely to participate relied on mnemonic techniques that can be learned, such as the "method of loci," an approach invented in ancient Greece based on the insight that recall can be vastly improved if one places whatever it is one wishes to remember at different locations inside an imagined "memory palace." To use this technique, you simply visualize a familiar building, house, or path in your mind's eye, populate each room or turn with an image of an object somehow associated with whatever it is you wish to remember.*

---

* In the bestselling 2011 book, *Moonwalking with Einstein,* the journalist Joshua Foer uses this technique to transform in less than a year from a neophyte observer, covering a memory competition for a magazine, into the eventual U.S. national champion—able to memorize an entire deck of cards in a minute and forty seconds

The Greeks and Romans knew that there is something about enlisting multiple senses, especially visual cues, that helps make memories stick. Indeed, some have suggested the momentous memory of Luria's famous modern subject, the Russian newspaper reporter Shereshevsky, was due to a powerful case of synesthesia, a rare condition in which the stimulation of one sense causes the recruitment of the others. Shereshevsky, for instance, often "saw" words and numbers when he heard them: One was a "proud, well-built man," two was a "high-spirited woman," and six was "a man with a swollen foot."

But the skills needed to learn the adept use of these techniques do not, it turns out, require good memory genetics. Tully knows because his people have brought eight memory contestants back for three days of extensive tests. Interestingly, the memory athletes' biggest strengths appeared to be not in memory, but in the areas of attentional control and working memory.

The XMT competitors that Dart's memory experts are truly hoping to find are those who have perhaps never heard of the method of loci or any of the mnemonic techniques used by most of the people up on the stage. They are looking for competitors who perhaps don't even know how good they are.

"The memory competitors use a very specific method and they know they're doing it—it's not a subconscious process," says Mary Pyc, a cognitive psychologist at Dart who helps coordinate the tournament and the testing of the competitors. "But we believe that there are people in the general population that have naturally occurring exceptional memory abilities. Perhaps some have decided to compete.

"The performance of people that have naturally occurring exceptional memory abilities will look like the people who did the training," she says. "But they might say something like 'I don't know. The information just sticks with me.' Maybe they don't need to do the training. That's where the genetics may come into this."

Though most of those tested relied on the ancient tricks of the trade and did not have the kind of naturally excellent memory ability that the team was looking for, Pyc says she did identify a couple of people she plans to test further.

XMT is just one tiny part of the sprawling effort to scour the globe for individuals with exceptional memory and eventually to study their genes. In addition to memory champions, Tully and his collaborators have begun testing *Jeopardy!* champions, and are considering studying top crossword puzzle solvers and chess champions. The preliminary results from the *Jeopardy!* champions were more promising than those from XMT, because they just seemed to soak up lots of information, and were later able to easily access that information from long-term memory, Pyc says.

Tully has recruited help from academia to undertake this first global search phase of the plan. (The company has not yet begun identifying genes responsible for superior memorization, but is preparing to do so soon.) For the last several years Dart has provided about $250,000 a year to Henry "Roddy" Roediger III, an affable psychologist at Washington University in St. Louis who has spent his career researching the ways in which human beings access and recall memories and has been present at each XMT event. Among other things, Roediger is developing standard tests that can be used on subjects with different areas of expertise to identify superior memory. XMT contestants, *Jeopardy!* champs, and individuals identified in the general public may all have areas of strength. The key is designing tests that measure just the abilities Tully is hoping to find a way to augment.

The Roediger lab, Tully, and Pyc (a former Roediger postdoctoral researcher) are using a variety of methods to look for memory outliers in the general population. One of the first things the team did was create a voluntary test, place it online through Washington University, and promote it through social media. "Good with names and faces?" it read. "Test your face-name memory." Remarkably, more than sixty thousand volunteers eventually clicked on the link. Roediger and his team found thirty-five individuals who seemed to score the highest, implying that unless they cheated, they had superior memorization abilities. But it's just the beginning.

"We're going for five million people," Tully says, "because we need to find about one hundred people in order to have the sample size necessary to statistically identify genes in the genome that are correlated with that extreme memorization."

Tully leads me down a long, out-of-the-way hallway, stops in front of a nondescript doorway that would be easy to miss, and beckons me into a small, darkened room. This, he tells me, is "the fly room."

Dart's operations and reach are truly sprawling. Dart has also opened a second research facility in western China, three and a half hours by plane due west of Shanghai in "the middle of nowhere." Dart NeuroScience is helping pay for the research of collaborators at, among other places, New York University, the University of Edinburgh, the University of Alberta, and National Tsing Hua University, in Taiwan.

"The NIH," Tully notes, "only funds people to study dysfunction, so nobody's ever studied superior memorization. It's really cool to see the academic community is totally fascinated by this stuff."

But the surprising truth is that Dart's most potent weapon might be in this modest little corner of the building, and the modest little organisms that reside here—fruit flies. Several times during our tour, Tully has marveled at his good fortune, looking around at his dream facility, and uttered the same sentence he now repeats in that dark hallway half in pride, half in something like disbelief.

"I mean, I'm just a fly guy!"

And it's true. None of it would have been possible without them. Studying fruit flies might seem an indirect route to unlocking the secrets of human memory and scouring the globe for humans with the most remarkable recall. It might seem entirely irrelevant to the feats of mental athletes, Greek philosophers, and a nineteenth-century blind savant who played the piano at the White House.

But in fact, in the mid-1990s, using the tools of modern genetics, Tully pulled off a feat that to many of his peers seemed every bit as exciting as anything these men ever did. Tully demonstrated he could give fruit flies the capacity to form the functional equivalent of photographic memories.

To understand how Tully did this, how it might be relevant to his current efforts and why a savvy businessman is willing to commit up to $2 billion toward them, we have to first take a little detour back down the rabbit hole of what's known and what's suspected about how exactly new memories and associations are formed on the most basic level of the brain, at the level of molecules.

For centuries, philosophers and scientific theorists alike have pondered the associative nature of our conscious, episodic, long-term memories. Why is it that one idea or sensation can bring to mind another independent idea or sensation linked only through the coincidence of its simultaneous occurrence in time? How, for instance, does the smell of someone smoking a pipe remind me of my grandfather Mannie? Why do I instantly see him in my mind's eye wearing a bow tie, standing in a brightly lit New York City room at the antique, cherrywood credenza he used as a bar, pouring himself some scotch?

It is paradoxical: Our conscious memories are both the easiest to subjectively observe and describe, and at the same time among the most complex and difficult to biologically deconstruct. When we recall a single memory, we are often in a sense reexperiencing not just all the smells and sights, but also the sounds, feelings, and ideas that collaborated to create the experience of a single moment in time. This is a miracle of biology. All of the disparate sensations—the smell of pipe smoke, the contours of that bar, the white shag carpeting, how stuffy it used to be in that apartment, and the feelings of childlike reverence and affection I felt in my grandfather's presence—are stored in different parts of my brain. Yet somehow, through the profound process of remembrance, I can once again in a millisecond reactivate neurons in areas that collected the initial sensory impressions and cause them to fire together again just as they did at the moment that the memory first formed so many years ago.

We know from amnesiacs like Henry Molaison that somehow it's the structures of the temporal lobe, hippocampus, and amygdala that allow for the storage of these disparate sensations all together. You might recall (thanks to your hippocampus) that after H.M. lost these structures,* he could still hold information briefly in his mind, he could still remember memories from before his surgery, but it was as if the filing clerk responsible for encoding and placing new information in long-term storage had left the building. Once a new piece of information disappeared from Molaison's view, from his conscious mind, it was gone. He had lost the ability to store it and thus to retrieve it. What these structures do exactly is unknown. But somehow, it seems, they knit together neurons across the brain, and

---

* The idea that these sensory associations are encoded somehow in the physical structure of the brain and in the connections between distinct groups of neurons is so fundamental that it dates back at least to the late nineteenth century, when scientists began to accept the foundational theory of brain organization known as "the neuron doctrine." The neuron doctrine states that neurons are in fact autonomous units, connected by projections that meet at the "synapse" between two cells, and that neurons pass signals across these synapses to cause one another to fire. In the late nineteenth century, Spain's Santiago Ramón y Cajal, whom many regard as the father of modern neuroscience, was among the first to suggest that the "strength" of the physical connections between groups of neurons encoding two distinct thoughts or ideas determine the strength of the association, and thus the likelihood that one idea or stimulus will bring to mind another.

store the patterns with which they fire in response to a whole collection of different inputs in such a way that we can later call them up, reactivate all these neurons together at once, and reexperience a moment in time.

From the earliest days of modern neuroscience, investigators have suggested that these explicit associations (just like implicit memories) are encoded in the connections between the 100 billion individual neurons that make up the brain. When Donald Hebb later suggested that "neurons that fire together, wire together," in fact, he had in mind as his primary example explicit memories. Hebb suggested that the brain was essentially a powerful coincidence detector, and that the physical rules governing how connections between neurons form and strengthen are designed to reflect and record these coincidences—and certainly this makes sense when one ponders conscious memory. Disparate sensations connected only by the temporal proximity of their occurrence in the past tend to rise to our conscious awareness in tandem. When the smell of a pipe reminds me of Mannie, it's because neurons in the sensory areas of my brain that fire when I process that particular smell are somehow *physically* connected to the neurons that encode the deeply buried memories of my grandfather.*

How else could the signals in the olfactory areas of the brain—signals prompted by the actual smell of pipe smoke coming in through my nose—reach the areas of the brain containing visual memories of his maroon bow tie and the stuffiness of that room? How else but by passing signals across synapses could neurons associated with that pipe smell somehow cue the neurons associated

---

* Marcel Proust famously wrote a huge book about his childhood memories called, appropriately, *Remembrance of Things Past*. The cue that sent him on his journey down memory lane was the taste of a single small cake, a madeleine, which pulled enough episodic memories to fill seven volumes of work.

with the bow tie, neurons located elsewhere in the brain, to also fire?

It's no coincidence that the experiment many believed first provided solid biological evidence for Hebb's theories involved neurons in the areas of the brain thought to serve as the file clerk for memory. In a seminal experiment in 1973, Terje Lomo and Tim Bliss for the first time demonstrated a Hebb-like phenomena in the memory centers of rabbits that they called "long-term potentiation" (LTP). By applying a series of electrical stimuli to neurons leading to the rabbit's hippocampus, the brain structures destroyed in the brains of amnesiacs like H.M., Lomo and Bliss demonstrated they could significantly strengthen the connections for an extended period of time. This strengthening was reflected by an increased sensitivity to stimuli passed between the connected neurons that lasted from several hours to more than a day.

Though no one has yet definitively proven that the phenomenon of long-term potentiation is what underlies episodic memory (at least not definitively enough to rule out other factors), most agree it is likely the way that we form memories. As we have seen, LTP, or Hebbian learning, is present in all parts of the brain, but the neurons in the hippocampus are among the most sensitive to it. (That is also true of the reverse, "long-term depression"—memories can weaken, too.)

It's only in the last couple of decades or so, however, that neuroscientists have begun to tease apart the processes that actually underlie the strengthening of a connection between two neurons at their most basic level—at the level of *molecules*. And it's only by reducing the brain to its smallest constituent parts and teasing out what precisely is occurring on the molecular level to strengthen the synaptic connections that actually underlie memory that one can realistically begin to attempt what Tim Tully has set out to do: Hack the brain and create a "memorization pill."

Yet by 2002, it seemed that Tully and some of his scientific peers had in fact come so close to this goal that *Forbes* magazine commissioned a writer named Robert Langreth to profile these efforts for a cover story that captured the attention of the man who would eventually come to back Tully's quest, Kenneth Dart. Though media shy, and reportedly a ruthless finance shark, Dart is in one way not so different from many of the other characters we have met so far in this book—he was trained as an engineer. At the University of Michigan, Dart earned his undergrad degree in mechanical engineering and learned to look at machines as the sum of their disparate parts. It's easy to see why Tully's efforts might appeal to an engineer.

Certainly the way Langreth portrayed the hunt to reverse-engineer the molecular basis of human memory and modify it harked back to classic engineering races of yesteryear—such as the race to the moon, the battle to construct the world's tallest building, or the race to complete the first transcontinental railroad. Biotech firms, he wrote, were "tantalizingly close to unraveling the mysteries of memory," and an "intense scientific race" was under way "to devise the first effective memory-enhancing drug."

And indeed, at the time of Langreth's article, a number of big names had joined the race, including Merck, Johnson & Johnson, and GlaxoSmithKline. But in Langreth's telling, the main action came down to a battle between Tully and Watson's Helicon and another small company, Memory Pharmaceuticals. That company was headed by the man Langreth described as Tully's "arch-rival"—a man whose earlier research in many ways set the stage for the unfolding race: Eric Kandel, "the elder statesman of the field," a Columbia neurobiologist whose earlier work on memory had won him the 2000 Nobel Prize.

Of course, the day-to-day realities of this "scientific race" and the events that made it possible didn't exactly read "epic." While Tully's chosen vehicle in this battle to the scientific finish line was a modest fruit fly, Kandel had spent years studying an equally mundane creature—a lowly variety of sea snail known as *Aplysia,* beloved by researchers because of the humongous size of its brain cells.

Yet somehow, in the pages of *Forbes,* Langreth the wordsmith masterfully transformed a series of meticulous, dry, laboratory experiments into something sexy. In Kandel's petri dishes, the neurons of the sea slug engaged in "a subtle electrochemical mating dance that reinforces links between them."

"A short-term memory is like a one-night stand, held together by fleeting but intense surges in chemicals that bind cells together," Langreth wrote of what Kandel discovered. "The effect fades away a few minutes or hours later. Long-term memories are more like marriages, cemented in place for years with new proteins that reinforce the synapses connecting the cells."

To engineer and characterize these partnerships, tawdry and fleeting or strong and lasting, Kandel had developed a powerful experimental paradigm. He created simple circuits of sea slug neurons in petri dishes, and set out to study how simple memories are formed by gauging one of the slug's basic reflexes: the withdrawal of its gill at a perceived threat. One of the neurons was a sensory neuron, to receive the equivalent of a shock; a second was a motor neuron needed to withdraw the gill. Kandel wanted to know the exact molecular level changes that occurred every time he applied the equivalent of a painful shock to the gill.

In nature, Kandel knew, the application of a shock causes the sensory neuron to release the biochemical messenger serotonin, which in turn sets off the molecular level cascade Kandel was interested in understanding. So, to simplify the experiment, Kandel applied single pulses of the serotonin directly to the ensemble of neurons, rather than an actual electric shock. Meanwhile, he recorded how electrical potentials changed with each subsequent pulse, and mapped the changes in the cellular architecture at the synapse that seemed to make any increase in electric potential possible. In his book *In Search of Memory,* Kandel goes into more detail about what he found. A single pulse of serotonin (the equivalent of a single shock) delivered to the sensory neuron within an ensemble of neurons, his team demonstrated, increased the presence of two specific chemical compounds at the synapse, called "cyclic AMP" and "protein kinase A," respectively, which *temporarily* increased the sensitivity of a neighboring cell to signals from its neighbor. This was Langreth's "one-night stand."

Long-term memories, on the other hand—the marriages of neurons—required an actual physical remodeling at the synapse that *permanently* increased the amount of neurotransmitters released in the area between the cells every time the sensory neuron fired. These changes were "structural" in nature—in other words, some piece of permanent infrastructure, in the form of new proteins, needed to be added at the synapse to facilitate this stronger signal.

It was only after *five* electric pulses in Kandel's neuronal ensemble that such a sacred union might occur; after five pulses, he showed, one of the two chemicals released at the synapse moved to the cellular command center in the nucleus of one of the neurons. Once there the presence of this chemical seemed to work a little bit of magic, switching on a crucial gene, which led to a chemical cascade that was like blaring the work horn at the end of lunch hour on a construction site: The activation of the gene caused an army of cellular construction engineers to put down their lunch pails, head to the synapse, and set to work building new connections between the two neurons. This took time.

Perhaps, Kandel would later suggest in *In Search of Memory,* the reason that electroconvulsive therapy or a big knock on the head disrupted long-term memory consolidation was that it somehow cleared the synaptic construction site of all these cellular engineers before they could finish their work.

It was a key discovery by Kandel in 1990 that got the race for a memory drug started. Kandel had come to suspect the involvement of a specific protein in flipping on the genes and inducing that lunch hour horn to emit its signal. It was called "CREB" (cyclic AMP response element-binding protein). And, sure enough in 1990, when his team figured out a way to block CREB in the sea slug, the organism seemed to lose the ability to form new long-term memories after the five shocks.

The relevance to a memory drug was obvious. Might it be possible to find a way to artificially help CREB do its job? Eventually Kandel would suggest there were two *kinds* of molecules particularly relevant to CREB: one that turned the gene to make long-term connection on or up (called "CREB activator"), and one that turned it down or off (called "CREB repressor"). It was the ratio of CREB activator to CREB repressor in the nucleus, Kandel suggested, that determined the strength of a long-term memory.

In his book, Kandel hypothesized that in a highly emotional state, such as that brought about by a car crash, or turning on the television and learning two airplanes had crashed into the World Trade Center, the stress hormones McGaugh had singled out might set in motion a cascade of chemical processes that flooded the nucleus with molecules that reduced CREB repressor (the brakes on memory) and amped up the amount of CREB activator (the on-switch of memory).

"This might account for so-called flashbulb memories," Kandel wrote. "Memories of emotionally charged events that are recalled in vivid detail . . . as if a complete picture had been instantly and powerfully etched on the brain."

Similarly, Kandel continued, the exceptionally good memory exhibited by some people, like Solomon Shereshevsky, "might stem from genetic differences in CREB-repressor that limit [its] activity." He wrote: "Although long-term memory typically requires repeated, spaced training and intervals of rest, it occasionally occurs following a single exposure that is not emotionally charged."

It was a provocative and exciting hypothesis, because if the secret to our memory really did lie in the ratio of CREB activator and CREB repressor, there was no reason why we could not develop drugs to affect that ratio. There was no reason to think we couldn't all be like Shereshevsky. There was no reason to believe we couldn't develop a "memory pill."

This brings us back to Tully's fruit flies.

In his laboratory at Cold Spring, Tully had developed a method of testing memory in his fruit flies every bit as basic and powerful as Kandel's sea slug circuits. He exposed his insects to two different odors. One odor Tully paired with an unpleasant shock—whenever a fly smelled the odor, Tully would soon after crank up the electric juice and zap the little guy. (A standard training session involved twelve zaps.) At the same time, Tully trained his flies to learn that the second odor was benign. After he exposed the fly to that smell, Tully did not deliver an electrical zap. After exposing the fly to each pair of odors twelve times—with the zap and the no zap—Tully placed the fly at the bottom of a T-shaped maze. At the top of the T, the two different odors wafted out from the opposite sides. At the intersection, the fly had to decide which direction to go.

It seemed to take Tully's flies *ten training sessions (with twelve shocks a session)* before they learned to run away from the odor that had been paired with the shock, and instead head to the neutral odor on the other side of the T. Even then, their memories weren't perfect—they still sometimes went the wrong way and got zapped. But after ten trials they seemed to reach their highest level of performance.

After Kandel demonstrated it was possible to block long-term memories by blocking CREB, the race was on to take the pivotal next step—to demonstrate it might be possible to *enhance* memory by somehow enhancing CREB. By 1993, Kandel's lab was hard at work attempting to pull off this trick with his sea slugs. But Tully had a secret weapon that gave him an edge with his flies: Jerry Yin. Yin, a postdoctoral student in Tully's lab, had succeeded in genetically engineering fruit flies to produce more or less CREB. In one breed, Yin had inserted DNA that made the fly capable of producing more of the protein that when present acted as the memory on-switch (CREB activator)—that lunch-hour buzzer to get the cellular engineers working. In a second strain of fly, Yin had inserted DNA that allowed the fly to produce more of the protein that when present acted as the memory off-switch (CREB repressor) that sent all the workers back onto their lunch breaks.

These fly strains had an additional feature: Yin and Tully had genetically engineered them so the scientists could turn their specialized CREB genes on and off. When the genes were in their dormant state, the flies produced the normal ratio of proteins and thus developed with normal memory. When Yin and Tully exposed the flies to heat, however, the genes turned on, causing the flies to start making more CREB activator or more CREB repressor. This allowed the two researchers to turn on the two genes just hours before any training began—and then see what difference it made in the ability of their flies to form new memories.

In 1993, the team demonstrated that when they overexpressed the CREB repressor they could impair the synthesis of proteins necessary to form long-term memories. Short-term memory and learning, however, were not affected. A colleague named Alcino Silva, meanwhile, soon demonstrated the same effect when he studied mice that lacked the CREB-activating gene.

Next Tully and Yin went for the brass ring—attempting to do the opposite, and improve memory. Surprisingly, the initial results were disappointing—even when they overexpressed CREB activator, fly performance failed to improve. Soon after, Tully was sitting glumly in an airport lounge in New York's JFK Airport, waiting to catch a flight to Tokyo, shaking his head, and wondering what they could have done wrong. Suddenly he had an epiphany that caused him to bolt out of his chair and run to a pay phone.

Tully and Yin had been pretty explicit in their instructions to their technician Maria: She was to take both normal and mutant untrained fruit flies and train them by delivering the odors with and without the electrical zaps every fifteen minutes, for ten training sessions. Yet like the normal flies, when Maria placed the mutant flies in the T maze after their ten training sessions, they went the wrong way just as many times.

In that airport lounge, Tully got Maria on the phone and blurted out his epiphany—maybe Tully and Yin were asking the wrong question! Perhaps the genetically modified flies also had the same *ceiling* on performance in that maze as the normal flies. (In other words, they would always head down the wrong direction a certain number of times.) But what if the question wasn't how *many* times the mutant flies would choose to head in the right direction after a standard ten-session training protocol? What if the question was how *fast* they could learn to do so?

"Maria, try one training session!" he told her, explaining that she should pair the shock with the odor just once and then test to see whether the mutant flies made fewer mistakes than the normal flies.

When she did, the results were unequivocal. The memories were instant. The flies with the memory volume switch turned all the way up could learn in one exposure what normal flies could only learn in ten. The flies had a form of photographic recall. Tully and Yin had proven it was indeed possible to hack memory by targeting CREB.

One of the first people Tully informed of his remarkable experimental results was the Nobel Prize winner James Watson, a respected elder at Cold Spring Harbor. It was a Sunday night when Tully strolled into Watson's office to share the news.

"He leapt out of his chair—I mean hands over his head, feet off the ground," recalls Tully. "And he said, 'We're going to be rich!'"*

* "Now, some people may interpret that as personally rich, but I don't," Tully says. "I interpreted that response as 'Oh my God, we're finally going to get grant funding.' I choose to interpret it that way."

It had yet to be proven, however, whether CREB also played a role in *human* memory. Several weeks later, Tully came across an article in the scientific journal *Nature* that sent him running back into Watson's office. It was another Sunday evening.

Dutch researchers reported that they had identified a genetic defect in humans linked to a devastating brain disorder called "Rubinstein-Taybi syndrome" (RTS). The mutation was located in a gene that interacted with the CREB gene during memory formation. Was it possible that the mutation the Dutch had discovered in human RTS patients did the opposite of what Tully and Yin had done in their recent experiment, preventing memory formation?

"Jim! Maybe these patients are mentally impaired because the long-term-memory switch is broken!" Tully practically shouted.

There was, the two quickly agreed, only one way to find out. They needed to begin testing small-molecule drugs. If they could find a drug to penetrate into the brain and moderate the CREB switch, then they could prove its role in memory and treat the disorder. To do so, the two founded a company called Helicon Therapeutics, a small corporation backed by private equity investors that at one point grew to have as many as seventy employees. They had competition, since a herd of other renowned scientists also believed that the dawn of a memory pill was close at hand, among them Kandel.

In fact, by 2002 at least three of these companies had Nobel laureates attached, all of whom were boldly predicting an FDA-approved memory drug was just around the corner. Despite the hype, despite the *Forbes* cover story, these predictions proved overly optimistic.

Most of the companies eventually went out of business or, as Kandel's company did, quietly sold out to larger pharmaceutical companies who later abandoned or downscaled the efforts. Some scientists suggested that the mechanisms underlying memory might in fact be far more complex in humans than in Kandel's snails. In addition, the drugs affecting the pathways targeted by Kandel could also have unintended and devastating effects on other areas of the brain. But the largest obstacle may simply have been a lack of resources and patience.

"Almost every drug in development fails," the journalist Sue Halpern wrote in her 2008 book, *Can't Remember What I Forgot,* which chronicled the demise. "Of five thousand compounds, only five make it through the $500 million gauntlet of preclinical testing to human testing, and of these five, one, typically, gets the approval of the FDA."

Tully readily admits his private efforts likely would not have survived much longer had it not been for Dart. (Dart actually bought out the other investors from Helicon in 2012 and merged it with Dart NeuroScience.) But Tully also insists his optimism has not faded.

So far, Tully and his team have had six drugs in clinical studies. Two have been terminated for toxicity-related issues. Four remain in the pipeline. Tully is tight-lipped about exactly what kind of compounds he has created so far.

"Obviously, I can't tell you the details of some of those targets, but we continue to be focused on the CREB pathway and its role in long-term memory formation," he says.

Tully has big plans: He aims to continue to enter two new molecules every year into clinical studies aimed solely at improving memory.

In 2000, James McGaugh received a bizarre e-mail from a thirty-four-year-old woman who claimed to have a "memory problem." The problem was not that she forgot too much. The woman claimed she could remember every single day of her life dating back to the age of eleven.

"Whenever I see a date flash on the television . . . I automatically go back to that day and remember where I was and what I was doing," wrote the woman, Jill Price. "It is non-stop, uncontrollable, and totally exhausting."

McGaugh agreed to meet with Price on a Saturday. Before he did so, he carefully prepared to test if she could identify the big events occurring on particular days going back to the 1970s (when she claimed her recollections began), and also to test the opposite, whether she could name the dates associated with important historical events.

When she arrived, McGaugh was amazed to find that Price was able to name the date Elvis died. She could tell McGaugh the day the *Dallas* episode in which J.R. was shot aired, and the date of the Rodney King beating in Los Angeles. Price even discovered a mistake in the reference book McGaugh used to prepare his list: It was on November 4, 1979, that Iranian students stormed the U.S. Embassy in Tehran, not November 5, as his reference book incorrectly stated, she insisted. McGaugh looked it up online. Price was right.

With further digging, McGaugh realized that the cues Price used to retrieve these memories were always autobiographical in a nature. She recalled that Bing Crosby died on a golf course in Spain on October 14, 1977 (a Friday), for instance, because she recalled hearing the news on the radio in her mother's car on the way to a soccer game. In the months that followed, as McGaugh began to probe the limits of Price's memory, he realized that in some areas it was no better than that of the rest of us. In order to be exceptionally strong, Price's memories had to be tied to some autobiographical event. McGaugh eventually named Price's condition "highly superior autobiographical memory" (HSAM).

Since publishing an initial paper on Price, McGaugh and his collaborators have identified roughly fifty-five more individuals who have the condition. Surprisingly, people with HSAM also display many of the symptoms associated with obsessive-compulsive disorder (OCD) in other areas of their lives, leading some to speculate that perhaps they are somehow subconsciously rehearsing the factoids they learn, creating almost implicit memories of detailed information. Tim Tully has already largely concluded that most of these individuals do not hold the key to the kind of memory he is seeking. But that doesn't mean this group has nothing to teach us. His collaborators are hoping they might find some with exceptional memory of the sort they seek in the pool of HSAM subjects identified by McGaugh.

Tully's collaborator Roediger believes he may already have found one. Through McGaugh, Roediger met Jake Hausler, that remarkable kid who memorized the inspection stickers on the cars in his New Jersey neighborhood when he was just a toddler. Roediger has met with Jake several times. And though McGaugh has diagnosed Jake with HSAM, Roediger believes there is more going on, though he is certain that Jake is not autistic.

"Not many four-years-olds can memorize inspection stickers," Roediger says.

By the time Roediger and I have this conversation, that precocious kindergartner has grown into an outgoing, thirteen-year-old boy with coppery brown hair and a broad, toothy smile. Today Jake collects baseball cards and likes geography. I decide I'd like to experience this firsthand, even if just by phone.

When I finally get the chance to speak with Jake myself, he and his mother, Sari, are sitting in the living room of their current home in St. Louis.

"Jake, why don't you ask Adam to tell you his birthday?" Sari suggests.

"When's your birthday?" Jake asks me.

"My birthday is August first," I say. "Why?"

"Oh that's easy," Jake says. "That was the day I went to the Detroit Tigers game!"

On August 1, 2014, Jake was visiting Detroit for the first time with his family, he explains. It was a Friday, a bit cloudy in midtown, but comfortable in the high 70s.

"We went to Greektown, a Greek neighborhood of Detroit, and had great food," Jake says. "Flaming cheese. It smells like gasoline."

I'm impressed. But not yet convinced. After all, it was just a few months ago. How about other August firsts? I ask him. Can you tell me about them?

"Well, there was 2012," Jake offers. "That was a bad one. I went to summer camp. At lunch I didn't feel good and I said I didn't feel so good. I said I feel like I may throw up and I felt dehydrated and surprisingly kind of congested. It was ninety-six that day."

Then Jake mentions Francis Scott Key, who wrote "The Star-Spangled Banner." He shares my birthday, August 1, Jake tells me. But he was born in 1779.

Like Price, Jake has found that having superior recall is not always easy.

"I really like having it but I can remember bad things that happened to me," Jake says. "And the bad things usually stick out more for some reason."

Like Friday, September 16, 2011, the day he accidently pushed a kid off a climber in the playground and got sent to the detention room. Now he has to remember it—the event itself and the way he felt—every day. For the rest of his life.

It's hard to speak with someone like Jake Hausler and not come away optimistic about Tully's quest. And with millions of baby boomers approaching senility, there's little question that such a product would create demand like few drugs we have seen before. I, for one, would love to stop worrying about what it means when I lose my car keys.

CHAPTER 8

THE SURGEON CONDUCTING THE SYMPHONY

DBS AND THE POWER OF ELECTRICITY

One gorgeous August morning in 2004, at around eleven o'clock, thirty-year-old Liss Murphy rose from her desk at one of Chicago's top public relations firms and, without bothering to pack up her belongings, shut down her computer, or tell anyone where she was going, walked out of her office and disappeared from life.

She'd felt something coming on over the previous couple of weeks: a creeping anxiety, a sense that something wasn't quite right. At times she'd found herself crying, and she didn't know why. But it came and went quickly, as if she were catching the outer edges of a storm. That August morning was the day the storm finally arrived.

Today Liss Murphy, a trim, blond woman with attractive, angular features and a wide smile when she reveals it, has a very hard time using the word *depression* to describe what happened to her. Depression implies something gradual, like the slow but steady deflation of a balloon as the air leaks out one end.

Her mental illness, when it hit in the summer of her thirtieth year, she recalls as a "fast and furious thing"—a spontaneous emotional tornado that appeared out of nowhere, picked her up, and sucked the life-force out of her, leaving her with "no thoughts," "no feelings," a space that was somehow both "empty" and profoundly heavy at the same time.

Liss Murphy never returned to her office. She never even picked up her things.

"It wasn't just a psychological breakdown," she says. "It was mental, physical, physiological; everything just crashed. I lost an incredible amount of weight, never left the house. I was mute for two years. I couldn't function. Everything was just an extreme effort. It was like when they put one those lead vests on you before they take your X-rays, like I was wearing this huge weight."

Within just a few months, she had left Chicago, moved back to her hometown of Boston, and begun building a medical resume that would grow to seven pages. She tried selective serotonin reuptake inhibitors (SSRIs), such as Prozac and Paxil, and talk therapy. She took monoamine oxidase inhibitors (MAOIs) and tricyclics (Nardil, Elavil, doxepin). She tried stimulants and "adjunctive therapy," different combinations of a wide array of medications meant to modify one another. Finally, she agreed to allow doctors to strap her down and give her electroconvulsive therapy—not once, but thirty times—at McClean Hospital. Even that couldn't jar her back into life. Nothing seemed to work.

Then, in May 2004, a psychiatrist named Darin Dougherty told Liss that he was looking for a volunteer to undergo an experimental procedure called "deep brain stimulation" (DBS). The surgery was not for the faint of heart. In an initial surgery, Dougherty explained, his colleague, a neurosurgeon named Emad Eskandar, would drill two dime-size holes in the top of Murphy's skull and sink electrodes 42 centimeters long and about 7 centimeters deep into the gray matter of her brain.

Then Liss would return for a second surgery. This time Eskandar would create a pocket under the skin in her chest, implant a device incorporating a battery and pulse generator into this newly created space, and run a wire under her skin up to her skull, connecting it with the electrodes. When turned on, the device would emit an electrical current that would stimulate the neural fibers carrying signals from primitive brain areas associated with motivation and emotion to the frontal lobes.

Dougherty believed something had gone fundamentally awry in the workings of Liss Murphy's brain. By hacking into the right part of it, by speaking to it in its own language of electricity, perhaps there was a chance to set things right. Of course, Dougherty couldn't promise Liss that this would solve her problems. By then DBS had been used for more than a decade to treat severe forms of Parkinson's disease, helping to alleviate tremors, stiffness, slowed movement, and problems walking. But Murphy would be Massachusetts General Hospital's first depression patient to have the procedure, and one of the first in the world, in fact.

There was, however, a good reason to believe it just might work. Eskandar had performed the surgery on a number of Dougherty's patients with obsessive-compulsive disorder (OCD), many of whom also happened to be profoundly depressed. And in many of the cases, something wholly unexpected had transpired—the therapy didn't just cure their OCD. It lifted them out of their depression.

Liss didn't take long to decide. She wanted her life back. Yes, she told Dougherty, she was willing to try.

As we've seen, neuroscientists often compare the firing of neurons in concert or in sequence to a "symphony" of many instruments playing together, producing a whole greater than its parts—a symphony that allows us to think, to feel, and to move. For decades, much of neuroscience—and many of the characters we have met so far on this journey—have been focused on listening in and making sense of this music.

But what Dougherty and Eskandar were proposing to do with Murphy was something else entirely. Dougherty and his colleagues weren't content simply to sit outside the doors and listen in on the various instruments of that neural symphony playing together. They wanted to open up the doors of the symphony hall, stride right up to the conductor's podium, and attempt to redirect the orchestra.

Many believe interventions like DBS and other techniques that allow doctors to change the neural activation patterns of the brain—to, in essence, conduct the brain's symphony of neurons—have the potential to revolutionize medicine. Existing drugs for brain disorders are often ineffective and frequently produce troublesome side effects. One reason is that drugs alter the chemistry of the entire brain, not just the area of interest, thus modulating the behavior of otherwise healthy neurons. With electrical stimulation, on the other hand, doctors can target discrete populations of neurons, confining the treatment to small, isolated areas of the brain that are causing the problems. Because they are doing so, they can concentrate the full force of the technology on these areas of interest.

"DBS allows us to go into the actual circuit that we know is involved in a condition, and we're stimulating it and making it fire or not fire in the way that we want it to," says Dougherty. "It's night and day in terms of the robustness."

Theoretically, there is no reason such techniques need to be confined to medical conditions. A number of experiments have already demonstrated DBS can be used to potentially modulate awareness and improve performance on a wide array of cognitive tasks—even to accomplish what Tim Tully has been attempting so hard to produce with a pill: improve memory. The results have proven so promising, in fact, that the military is funding a wide array of less invasive techniques that attempt to do the same thing without brain surgery.

While working on this book, I myself in fact allowed a researcher at George Mason University to attach an alcohol-soaked sponge to the top of my skull, turn a dial on a black box, and send electricity pulsing through the right side of my brain's parietal cortex. (I felt a slight burn, nothing more.) The technique is called "transcranial direct current stimulation" (TDCS), and the researchers wanted to know if it could enhance my performance on a task that required short-term memory by potentiating the neurons in my brain. (They concluded it worked in others, though I never learned if it made me smarter. I couldn't tell.)

Alvaro Pascual-Leone, the Harvard neuroscientist who performed brain scans on Pat Fletcher, is experimenting with ways to use TDCS to accelerate neuroplasticity.

For now the evidence that these less invasive techniques actually work is still in the process of accumulating, which means that for the foreseeable future the most exciting scientific developments, the hardest ones to dismiss, are likely to continue to emerge from experiments conducted on individuals who have actually had electrodes implanted directly into their naked cortex. This direct access to neurons allows neurosurgeons to control electricity without the dampening and scattering effects of a skull in between, and is thus far more effective.*

---

* Targeted brain stimulation could be used, theoretically, to enhance a wide array of mental functions. DBS could, for instance, be used to accelerate learning by making specific groups of neurons more likely to fire in tandem, and thus fuse together. And in fact, Eskandar has already demonstrated that he can dramatically enhance the rate at which monkeys can master a visual learning task by using precisely this technique. Itzhak Fried, a professor of neurosurgery at the David Geffen School of Medicine at the University of California, Los Angeles, has demonstrated the ability to enhance memory formation in human epilepsy patients simply by stimulating the areas involved in coding new memories. Fried is now one of the leaders in a DARPA-funded project aimed at further developing these as a therapeutic technique. The program is called Restoring Active Memory (RAM),

That is why, one summer day a few months after sitting in that room with the researcher from George Mason, I packed a bag and flew up to Boston to meet with the man who operated on Liss Murphy, the neurosurgeon Emad Eskandar. Despite my benign experience with TDCS, I wasn't about to volunteer to have electrodes implanted on my bare brain. But after clearing it with Eskandar, I did intend to stand behind the surgeon and watch from over his shoulder as he implanted them into someone else.

DBS was born in a French operating room in 1987, when neurosurgeon Alim-Louis Benabid made a fortuitous discovery while preparing to operate on a patient suffering from uncontrollable trembling. For decades, the last-ditch technique for such patients had been extreme but often quite effective: Brain surgeons simply drilled holes into the skull and removed the areas of the brain thought to be causing the problem. The approach was sometimes used for other movement disorders, as well as severe epilepsy and some mental illnesses. That day in 1987, Benabid planned to remove a piece of his patient's thalamus, a walnut-shaped structure deep in the brain. By destroying or "lesioning" part of the tissue, he intended to cut out the source of the stray electrical impulses flying down the peripheral nerve fibers of the body and causing his patient's hand to shake.

Brain surgery of any sort, of course, is a high-stakes proposition. Miscalculations can cause paralysis, blindness, even death. To avoid

---

and is aimed at helping individuals who suffer from traumatic brain injury. Fried will be collecting data from epilepsy patients. This and other data will be used to design and build devices that might improve memory formation. Fried's team will focus on the entorhinal cortex, according to the Los Angeles Times. A second team at the University of Pennsylvania plans to explore the contributions of other parts of the brain to memory, including the frontal, temporal, and parietal cortices.

surprises, Benabid took a common surgical precaution: He kept his patient awake in the operating room, which is possible because there are no neural pain receptors in the brain. He inserted an electrical probe into the part of the brain he intended to remove. Then he delivered a pulse and watched the patient closely to make sure the stimulation had no unanticipated effects. It's a technique neurosurgeons have been using for well over half a century to verify that the area they are about to remove does not serve an essential function; the small current in the electrode causes the neurons around it to fire, revealing what, if any, role they play in bodily processes.

By 1987, neuroscientists had developed a protocol that Benabid fortunately decided to ignore. Instead of stimulating the brain of his patient at a frequency of 50 Hz, he turned the knob up to almost 100 Hz. When he applied the electrode to his target, something unexpected occurred. The patient's hand stopped shaking—for the first time in years. When Benabid turned off the current, the shaking resumed. When he turned it back on, it stopped again. Stimulating at high frequency, he realized, somehow quieted the troublesome signals.

As he began attempting to understand this mysterious technique—and develop the most effective ways to use it—Benabid was helped by a parallel discovery made by an Emory University neuroscientist, Mahlon DeLong. For much of the twentieth century, most scientists believed that signals from all across the sensory and motor areas of the cortex converged in the deeper brain areas associated with motivation, reward, and movement, where they were somehow mixed together and evaluated. At the end of this process, mysteriously, a signal would emerge from this area that caused a person to respond appropriately—to move their hand away from a hot stove for instance, or adjust their step to climb a set of stairs.

But in the early 1980s, DeLong discovered something curious while recording and analyzing the brain activation patterns in monkeys. DeLong, whose target was the basil ganglia, found that as monkeys performed tasks requiring different kinds of movement, there was no "convergence," or mixing together of different sensory signals from the cortex, as science would have predicted. Instead, the sensory signals that arrived in the basil ganglia from different parts of the body stayed segregated from one another, traveling in discrete "circuits" or networks that ran in parallel. The area of cortical real estate studied so assiduously by Mike Merzenich that was devoted to processing the feeling of touching something with a fingertip, for instance, was connected to a corresponding localized area devoted to touch in the fingertip in the basil ganglia, which then connected to a third way station for processing this feeling in the thalamus—and then fed back into the cortex. At each of these four way stations, inputs from other areas of the brain fed new information into these discrete circuits, but the primary signal remained the same, moving along the same fingertip "circuit."

DeLong went further than simply proposing that there was continued segregation of sensory and motor processing as this information moved through the brain. He argued that we used this same system of discrete circuits to perform higher-level functions, like cognition and the processing of emotions. In fact, there were discrete "parallel" or "modular" brain circuits running from the cortex to the basil ganglia or its neighboring structures to the thalamus and back to the cortex, he suggested, for virtually all functions the brain performed.

"These loops seem to be key to understanding almost every aspect of behavior and how learning and maladaptive behaviors can result in disturbances," DeLong says. "It's as though in evolution as new functions were acquired, new modules were added to this very basic primitive organization on top of one another."

This theoretical finding had very real implications for medicine. DeLong began to tease out the separate circuits, identifying motor and nonmotor pathways. And he identified one crucial structure, or "way station," in the basal ganglia in particular that seemed to play a key role in Parkinson's, called the "subthalamic nucleus." In monkeys with Parkinson's disease, this area was unusually and extremely overactive. When DeLong immobilized the area using a drug, however, the symptoms disappeared. In at least some cases, he had shown, brain diseases were diseases of circuits—and if you treated the right circuit, you could get rid of the symptoms.

When Benabid read DeLong's paper, he realized he'd found the ideal proving ground, the perfect test target, for his new DBS technique. After publishing a watershed 1991 paper detailing his use of the technique to treat tremors on both sides of the body for patients suffering not just from essential tremor and Parkinson's disease, he followed up with a second landmark paper demonstrating he could alleviate many of the other debilitating symptoms caused in Parkinson's, including slowed movement and muscle rigidity, by following DeLong's lead and stimulating the nearby subthalamic nucleus.

Even today, some thirty years after the discoveries of Benabid and DeLong, however, debate continues to rage about what exactly happens and why when neurosurgeons insert electrodes into the gray matter of the brain and deliver this heavy blast of electricity. For years many neuroscientists and neurologists believed that stimulating the neurons suppressed the abnormal activity in a specific circuit, perhaps by somehow tiring it out. This would explain why DBS worked so well to quiet the uncontrollable trembling seen in patients suffering from Parkinson's disease, and why it has proven effective in quieting the neuronal chatter that seems to operate on an endless loop in patients inflicted with OCD.

Over the last decade or so in animal studies, however, neuroscientists have more precisely measured neuronal output and found that DBS at the frequencies used seems, on the contrary, to stimulate neuronal activity. Philip Starr, a neurosurgeon at the University of California San Francisco (UCSF) who specializes in movement disorders, has articulated a leading theory: He believes that DBS works by "desynchronizing" firing patterns within circuits of neurons.

Just like energy moving through the ocean, electrical signals passing through the circuits of the brain travel in waves. And as in an ocean storm, a big wave moving at the right speed can subsume all the little waves in its path. In Parkinson's, abnormal activity builds on itself, creating pathological waves of activity that gain control of the circuit, drowning out all other activity. DBS breaks these waves up again, allowing the circuit to unlock and smaller signals to get through.

Whatever the cause, the FDA approved DBS for use in tremors in 1997, and for Parkinson's in 1999, and it has now been used on tens of thousands of patients to calm the symptoms. It was only a matter of time before researchers began to consider how they might extend this theory of circuits and this technique to treat other brain disorders—in particular, intractable mental illnesses.

The balmy July day I walk through the polished, glass lobby of Massachusetts General Hospital is a gorgeous one, much like that day in Chicago almost ten years ago when Liss Murphy walked out of her office, never to return.

Murphy's experience with brain surgery, of course, has long since concluded. But when I arrive in Emad Eskandar's office and announce I am ready to watch today's operation, he assures me the procedure has changed little since then. After equipping me with my own pair of surgical scrubs and a face mask, Eskandar warns me that he moves fast. It will be up to me to keep up.

Eskandar leads me down a dizzying warren of brightly lit hallways and elevators, past slow-moving relatives looking dazed and lost, through doorways that require key cards, past anesthetized patients lying bleary-eyed on stretchers, pushed by fast-walking nurses in clogs. We stop abruptly in front of a door leading to the operating room.

Today's patient is laid out on a stretcher wearing a festive dark blue shade of fingernail polish. Like Liss Murphy, this patient is desperate, having failed to respond to all other treatment options. Rather than depression, this patient's personal tormentor, however, is obsessive-compulsive disorder. Though often referenced casually in popular culture—"I'm sorry, I'm a little OCD. I arrange my shoe collection by color"—it's a condition that, in the case of Eskandar's and Dougherty's OCD patients, is far more debilitating. These patients often take three-hour showers. They spend eight hours cleaning their surroundings with bleach. They can't work. They have even been known to get stuck at the bathroom sink in their hotel room on appointment days, unable to stop washing their hands until someone comes to get them. One patient, Eskandar tells me, had a "thing" about mirrors. Prior to surgery, Eskandar had to cover up every single reflective surface on the way to—and inside—the operating room.

Of course, today's patient is currently lying peacefully anesthetized, unconscious, amid trays of shiny metallic scalpels and scissors, bathed in lights powerful enough to equip a tanning salon. The nurses have draped her in a white sheet. They have also shaved her head and, using clamps and screws, secured a sturdy, boxlike frame to her forehead and the sides of her skull. Each arm of the frame is etched with the tiny numbers of a ruler, down to the millimeter. The numbers will allow Eskandar to precisely line up the hollow metal leads he plans to press through his patient's cortex and into the center of her brain, following a straight route to his target.

First, however, the neurosurgeon needs to map that route. Eskandar takes a seat nearby in scrubs, a surgical mask rakishly pushed up onto his blue surgical cap, and moves a mouse pointer over a spot at the center of one of four images displayed on a monitor depicting the patient's brain. Each image is taken from a different angle. This is the result of all of those efforts to map the brain using the latest imaging technology. Eskandar has a target where he believes a well-placed burst of electrical activity will emanate out through the circuit involved in OCD and hopefully liberate his patient from the condition that has ruined her life.

"This is perfect—you want to be here," he tells a junior surgeon, pointing at a spot on one of the images. "That's your entry point."

Along with Dougherty, Eskandar today directs Mass General's Division of Neurotherapeutics, the nation's busiest center for psychiatric surgical treatment. And in recent years, the duo have only continued farther down the path on which, several years ago, they first encountered Liss Murphy.

Today, as part of President Barack Obama's BRAIN Initiative, the two are coleading a team of doctors, scientists, and engineers participating in a five-year, $30 million effort to develop the next generation of DBS and use it to treat severe psychiatric disorders, most of which have been considered far too complex and mysterious for any DBS system currently on the market. Not just depression, not just OCD, but conditions like schizophrenia, post-traumatic stress disorder, traumatic brain injuries, and anxiety.

All of these conditions are characterized by unpredictable changes in the brain that lead to intermittent episodes. And to truly and consistently tame these complex disorders, Eskandar and Dougherty have come to believe, it will require a new kind of device, capable not just of continuously stimulating the brain in a couple of places, as today's devices do, but also of monitoring brain activity in real time, detecting anomalies and then responding with intermittent bursts of electricity. Here the same kind of pattern recognition technologies we have seen in so many other arenas will once again play a crucial role.

In many cases, neuroscientists have not yet identified what these neural anomalies even look like. Thus, as part of the project, Dougherty and Eskandar are attempting to identify how the brains of people suffering from these disorders differ from those of healthy individuals. Then they will have to figure out what kind of electrical stimulation patterns might be used to fix them.

"We're aiming for something ridiculously ambitious," Eskandar acknowledges.

It is not, however, a pipe dream. Engineers across the Charles River at Draper Laboratory are working closely with Dougherty and Eskandar to develop the needed hardware. In the 1960s Draper's engineers famously played a key role in one of the twentieth century's seminal scientific achievements—helping design the guidance systems and instruments used on the Apollo 11 moon landing mission in 1969. Today the laboratory's top minds are training their computational and design expertise inward to a twenty-first-century frontier, the human brain. They have already developed a prototype of a device that combines DBS with many of the same powerful tools we have learned about in previous chapters—real-time brain imaging and recording technologies, pattern-recognition software, wireless technologies and exploding computer processing power. In some ways, such an upgrade is long overdue. The device to be implanted in today's OCD patient, Eskandar says, uses technology that has been around for decades—which is part of the reason why the veteran neurosurgeon is so certain that he and other clinicians have only scratched the surface of what might be possible.

"Think about what's happened over the past twenty years in terms of miniaturization and Moore's law and everything," Eskandar says. "You have this device that came out in the nineties. When it was designed in the 1980s, I didn't even have a cell phone."

Unlike the device Eskandar plans to use for today's procedure and the one he used for Murphy—which can only deliver electricity—Draper's prototype of a DBS system will also be able to monitor and record signals. And it will do so from as many as 320 electrodes—including multiple groups of sensors placed on the outer layer of the brain. The device will then use pattern recognition software to detect anomalous activity associated with pathological mental states and stimulate areas of the brain in response.

Instead of a bulky processor implanted in a patient's chest or abdomen, the device will consist of a miniaturized central hub, smaller than a cell phone, with an integrated battery. The whole thing will be compact enough to fit snugly on the back of the skull. The skull hub will attach to as many as five ceramic and titanium electronic satellites that are small enough to fit into dime-size burr holes drilled into the top of the skull. Each of these satellites will collect and relay the data from the electrodes that will be connected to the sensors or the leads deep in the brain. The Draper engineers are in the process of fabricating a miniaturized version of the device, which they hope to test out in humans in the months ahead, and which could theoretically remain implanted for years on end.

By comparison the device Eskandar is about to implant in the brain of his current patient is rudimentary indeed. The current device can be turned on and off by Dougherty, and the strength of stimulation can be modulated. Each of the two electrodes has four "leads" capable of stimulating at different points, for a total of eight (compared to 320). It is not capable of sensing dynamic states, and it certainly is not capable on its own of responding to them. Even so, for today's OCD patient it could be life changing.

In the operating room, Eskandar stands over his patient's naked scalp and marks his entry points with a Sharpie. Then he snaps an attachment onto the metal frame encasing the patient's head, adjusts the angle to line up the numbers, and lets the assemblage of nurses, residents, and other observers know he is ready. Within a few minutes, he has opened up two boreholes in the patient's skull with a drill and used the head rig to pilot two long, hollow metal tubes down through the outer layers of her brain and into the middle of the gray matter. Into the tubes he slides a pair of thin electrodes that will be connected to the device he plans to implant later. Then he removes the tubes, stitches the electrode leads into the scalp using silk threads, and fills the boreholes with fast-setting cement.

By now, this part of the surgery is almost routine. Eskandar has implanted electrodes like these in scores of OCD patients. He was, in fact, among the first neurosurgeons to begin performing the intervention experimentally, long before it was approved for widespread use by the FDA in 2009. It was an opportunity he had been hoping for ever since he decided to attend medical school.

Eskandar had excelled at math and physics in high school and entered the University of Nebraska intent on becoming a chemical engineer. But that changed when he got a night job working in a psychiatric institution, supervising patients experiencing acute psychotic breaks. The patients he met made a profound impression. There was the mathematics professor with a Ph.D. from Northwestern, hopelessly confused by his own delusions. Eskandar also recalls a disheveled guy his own age, who heard voices in Van Halen songs ("Can't you hear it?" he kept asking Eskandar. "They're saying 'Book 'em, Danno!'"); one day the kid hopped the fence during an outdoor recreation period when Eskandar wasn't paying attention and went AWOL, much to the future surgeon's embarrassment and consternation. The police found the patient a couple of hours later, standing in the middle of the freeway directing traffic with a fork.

Eskandar was fascinated by the magnitude of these delusions and amazed by how little doctors understood about mental illness. "It was a very different feel than a regular hospital," he recalls. "It was like, 'Does anybody really know what's going on?'" He applied to medical school hoping to unlock the mysteries of the brain. After a stint doing brain research at the National Institutes of Health, he earned a residency at Mass General just as the FDA was approving the first use of deep brain stimulation for movement disorders. Having entertained and kept watch over patients with brain disorders just a few years earlier, he now found himself operating on them, and in the process he got the opportunity to measure their neural activity and join the hunt for the causes of such bizarre behavior. He stayed at Mass General after his residency ended.

Within just a couple of years he was on the cutting edge of efforts to extend the surgery to psychiatric conditions. As brain surgeons considered new candidate conditions for DBS, OCD was the logical early prospect.

Just like Parkinson's, OCD is characterized by hyperactivity—errant firing patterns—in a loop running through the three main neural way stations identified by DeLong: the cortex, striatum, thalamus, and back to the cortex. Like Parkinson's, this riot of neuronal activity stands out any time you measure the brain, even during a period when an individual is at total rest, when in healthy individuals the brain would usually appear relatively silent. When looking at the brains of patients with Parkinson's or OCD, it appears as if somewhere in their respective loops a record is stuck in a groove, playing the same refrain over and over—or in this case resending the same neural signals over and over and over again, causing that distinct Parkinsonian shaking, or, in the case of OCD, the maddening compulsion to wash one's hand over and over and over again.

In the brains of those afflicted with OCD the repeating neural signal originated in a loop that began in an area of the frontal cortex called the "orbital frontal cortex," which neuroscientists had identified as playing a key role in emotion, reward, and decision making. From the frontal cortex, the loop traveled to the striatum, on to an area called the "dorsomedial thalamus," and back to the cortex. The hyperactivity looked similar to that seen in Parkinson's, but since it was present in a different circuit, one involved in cognition, it led to obsessions and compulsions rather than the physical tremors that occurred in Parkinson's.

The procedure was an ideal early therapy candidate for Eskandar and Dougherty, who for the past fifteen years have codirected one of the largest centers for surgical psychiatric interventions in the world. But within months of treating their first OCD patients, Eskandar and Dougherty were struck not just by how effective the procedure was on OCD, but by something even more unexpected. OCD and depression are highly comorbid—meaning many patients who suffer from OCD are also depressed. And almost 90 percent of the OCD patients who also had depression reported soon after the trial began that their depression seemed to be lifting, too. Dougherty and Eskandar soon won approval to conduct a small trial that used DBS to treat patients without OCD but suffering from depression. And then Darin Dougherty met Liss Murphy.

Though Murphy doesn't remember much about her surgeries—she was out for most of them—she remembers the moment several weeks later when she returned to Dr. Dougherty's office and the psychiatrist first turned on the electricity in her brain stimulation device. Liss insists she couldn't feel the difference that day. To her, life remained the same sluggish, gray slog. But in retrospect she has come to believe beyond the shadow of a doubt that it was in that moment that everything began to change. And for those clinicians who were there that day, the effect was impossible to miss.

Alice Flaherty, a colleague of Dougherty's at Mass General who came to know Murphy well, recalls Liss sitting hunched over, avoiding eye contact. Then Dougherty turned on the device, and cranked it up to 7, and something amazing happened. All of a sudden Murphy straightened up. She started asking questions. She met her doctor's gaze.

"Well, how do you feel?" Flaherty recalls asking her.

"I feel just as sad," Murphy responded. But the *way* she responded suggested something entirely different—there was a marked vibrancy, a "perkiness" in the tone that hadn't been there before, Flaherty insists.

"It fascinated me in terms of the relationship between action and emotion," Flaherty says. "We could see a huge difference. But we realized it kind of takes a while for the behavior to percolate in and have the body decide, 'Oh, I'm not depressed anymore or something.'"

Liss herself began to notice the change a few weeks later. She and her husband, Scott, had a shaggy Old English sheepdog named Ned. Before the surgery, the idea of taking Ned out for his morning walk seemed about as daunting as walking through a minefield. Every morning when the time came, Liss would call Scott at work and summon him home to take the dog out. But one day a few weeks after the implant, without hardly realizing she'd decided to do it or even considering why, Liss took the dog out herself. Soon she was taking walks with Ned every morning.

Then, four months after the surgery, Liss's grandmother died, and Liss volunteered to do something that would have been unthinkable just a few months earlier: She wrote a eulogy. Just a few months removed from almost total muteness, Liss stood up at the front of the service and delivered the heartfelt words to a small crowd of friends and family, touching upon her own journey and noting that her grandmother too had suffered through depression. A few months later, Liss returned to work part-time. The biggest payoff was yet to come. In 2012, Liss and Scott had a son.

For Liss, if there was any question of the need for the device, it was dispelled when an infection required doctors to shut it off for several months. Within days, her depression returned; but when the device was turned back on, she says, her experience was the exact opposite of that initial session in Dougherty's office that made such an impression on the doctors present. This time Liss Murphy was immediately aware of a powerful transformation.

"It was just a surge of warmth that rises through you, and I could tell it was on," she says. "I woke up the day after and it was a whole new world. The colors outside were brighter. My son and I went to the story hour. It had been months since just he and I had done anything. Everything was new again, and it was like I made it to the other side."

Yet in some ways, both Eskandar and Dougherty have come to believe, they may simply have gotten lucky with Liss Murphy.

That's because the problems that occur in depression, like most other psychological disorders, are not confined to one anatomical location. They are diseases of neural circuits and usually present complex arrays of symptoms, which might vary depending on which part or parts of the circuit are affected. This means there are different varieties of depression, and different varieties of patients with depression; each person might respond differently depending on where, when, and how the brain is stimulated.

This is something the two learned the hard way. Inspired by Murphy's success, the two clinicians expanded their explorations to other depression patients in the mid-2000s. The results, in some cases, were just as remarkable as those seen in Murphy, hinting at the potential the team is now attempting to realize. But in many other cases, the treatment was frustratingly ineffective.

By then, a parallel effort to use DBS against depression was already under way. In March 2003, Helen Mayberg, a neurologist then at the University of Toronto, had implanted a DBS device into a patient with depression, placing it in a separate brain area, a narrow structure called the "subgenual cingulate." She published a paper in the journal *Neuron* in 2005, a year before Murphy's operation, reporting results in six subjects. (She followed that up with a group of twenty, who are still being followed today.) Like Murphy, some of them had been virtually catatonic before the surgery but recovered.

Mayberg's initial success with DBS, along with the work of Eskandar and Dougherty's group, fed the widespread expectation that the device would soon win FDA approval for a condition affecting millions of Americans. Both groups had somewhere around a 50 percent response rate, with remission in a third of the cases, according to Dougherty. But the large trials the FDA mandated before the treatment could be approved required control groups to measure placebo effects. Experimenters implanted DBS devices in all the volunteers; then they randomly assigned half to a standard protocol of stimulation and the other half to a protocol in which the electrode is never turned on. After analyzing preliminary results, the FDA halted both trials around 2014.

"We ended up having a fairly high placebo effect," Dougherty acknowledges. "But it definitely worked in some people."

Eskandar and Dougherty had seen too many remarkable recoveries to discount the treatment. Mayberg also remains a staunch believer in the power of DBS to treat depression. But a more advanced device could mean far more precise interventions tailored to individual patients and, perhaps, an effective treatment for a larger group of people. It could also yield key insights into mental illness.

Though neuroscientists have learned a lot about how brain circuits are organized and how they function, it's rarely been possible to watch these circuits operate in real time for extended periods. But Eskandar and Dougherty say the technology they are designing and testing will open up that possibility. Already their group has begun to gather extensive data using existing brain-scanning technologies.*

---

* It's not hard to imagine, if you are a cynic or a medical ethicist, that brain stimulation devices might someday be used for some nefarious purposes. In the 1950s and 1960s, Yale neuroscientist Jose Manuel Rodriguez Delgado implanted radio-controlled electrodes into the brains of twenty-five human subjects at a Rhode Island mental hospital, most of whom were suffering from schizophrenia and epilepsy. Delgado demonstrated that he could elicit specific physical movements by stimulating the motor cortex, even when these patients tried to resist. He also could induce fear, lust, or euphoria. *Scientific American*'s John Horgan traces the demise of these kinds of scientific explorations to 1970, when two of Delgado's collaborators, Frank Ervin

---

Sitting in Eskandar's lab, I watch a rotating 3-D image of a translucent skull and the brain within it. Within the black-and-white brain, distinct neural activation patterns are highlighted in three different colors: turquoise, orange, and magenta. These colors mark the activity in various neural circuits associated with mental tasks. To create the images, Eskandar's colleagues used fMRI. The turquoise represents the brain activation patterns recorded from a healthy subject as he performed a specific task. The orange and magenta represent the activation patterns recorded from the brains of two psychiatric patients as they performed the same task. All three patterns appear different. Although the orange and magenta patients have both been diagnosed with major depression, each has an additional condition: One is suffering from PTSD and the other has generalized anxiety disorder.

"These disorders, by very definition, are constellations of symptoms," Dougherty says. That is why, he argues, a more precise treatment, better tailored for individual patients, could make all

---

and Vernon Mark, of Harvard Medical School, published a controversial book called *Violence and the Brain*. In it they suggested brain stimulation and lobotomies might be used to tamp down the violent tendencies of African Americans rioting in the cities. Two years later Robert Heath, a Tulane University psychiatrist, reported that he had implanted an electrode in the neural pleasure centers of a homosexual man, hired a female prostitute, and then turned on the stimulator in the hopes of changing the subject's sexual orientation. (It didn't work.) Congressional hearings followed, grants became far harder to obtain, and most researchers redirected their efforts at pharmacology. It's taken years for the field to recover. These incidents present a grim reminder of the potential for misuse in this area in particular. But it's hard to dismiss the potential good that DBS can offer after speaking with someone like Liss Murphy. And it seems that as memory has blunted the impact of these abuses, the epidemic of mental illness has finally overridden, at least for now, these concerns. We have another chance to get it right and use it for good.

the difference. "There is no depression spot," he says. "There is no PTSD spot. There is no borderline personality disorder spot."

Using the DBS system that's currently available, Eskandar explains as he points at the two depressed patients' brain patterns, the treatment strategy would be simply to turn on an electrode and stimulate the same area of the brain for both patients. The advanced DBS system Dougherty and Eskandar are developing with Draper, in contrast, will be able to sense abnormal patterns of brain activity in real time and stimulate whichever areas in the circuit are affected. They should adjust when new patterns crop up, applying a jolt of electricity in the right spot each time.

Eskandar once again calls my attention to the screen. The three brain scans we are looking at, he tells me, were all recorded while the patients performed a task that measured their ability to quiet the emotional areas of the brain and answer a question that required focus and mental clarity. Eskandar points to one of the depressed patients' brain activation patterns, explaining that it is the same pattern one usually finds in patients experiencing symptoms of PTSD. The emotion-driven part of the brain, called the "amygdala," is alight with activity. It is firing far more robustly than the amygdala of normal patients performing this same task. It's as if the emotional part of this patient's brain is screaming, drowning everything else out.

Imagine, Eskandar suggests, if we could simply override this reaction, manually activating and deactivating the appropriate areas. In fact, he has already attempted to demonstrate just that in a patient who had electrodes implanted in preparation for surgery to treat epilepsy. Eskandar and his team could turn up the patient's emotional response to a picture of a human face by stimulating the amygdala, and they were able to blunt that response by stimulating a different area, the dorsal anterior cingulate cortex.

The team hopes to design a whole host of new DBS treatments: The device's electrodes will be inserted at locations chosen according to each person's constellation of symptoms, and the particular abnormalities in the brain circuits will determine where the current will be activated. Eskandar is optimistic about the prospects for treating depression with these new tools. He also has high hopes for treating PTSD and generalized anxiety disorder. He even feels good about the possibilities for treating addiction, schizophrenia, and traumatic brain injury. But he acknowledges that some of the conditions he and Dougherty plan to target, such as borderline personality disorder, remain long shots. Even in the one psychiatric disorder for which DBS is FDA approved, OCD, the success rate still hovers around 50 percent—a stark reminder of the challenges that lie ahead.

Indeed, Eskandar and Dougherty are under no illusions. The human brain remains one of the most enigmatic and complex biological systems known. In many ways our efforts to understand it are still in their infancy. "I'm sure it's not going to work the first time, or probably even the third time," he says. "But eventually it's going to work. And we'll keep trying until we get it right."

CHAPTER 9

SUDDEN SAVANTS

UNLEASHING THE INNER MUSE

It's amazing to think that we may be on the verge of figuring out how to restore and augment the capacity of the human mind to record and replay the experience of individual moments in time, and that scientists are developing methods that might someday soon allow us to manually rewire the emotional circuits of the brain when they go awry.

But there is one final mental marvel that is in some ways even more elusive and fantastical to contemplate. Some scientists in recent years have launched ambitious efforts to understand and augment that most ineffable of human qualities, the capacity of the human brain for creative expression.

As this book comes to a close, a chapter on creativity seems an appropriate capstone. Some say creative expression is what distinguishes human beings from other species. Certainly it is among our most remarkable and powerful capabilities for expressing our humanity. A great song, painting, poem, or novel can work powerful magic—conjuring up visions or emotions from deep within us, making us feel connected to one another, and, at their very best, somehow causing us to feel more alive. Creativity is also among the most powerful tools we have to survive and evolve, individually and as a species.

You could say it's the spark of creativity that makes this book and all the science it contains possible. Creativity allowed Hugh Herr to look at his new prosthetic legs and visualize unusually long limbs perfect for the rock-climbing wall. Steve Badylak considered whether a piece of intestine might suffice as a canine heart valve. And what other source but this magical human trait could Mike Merzenich have drawn upon to dream up the experiments with monkeys and spinning disks to test his hypothesis on neuroplasticity?

Where does this profound mental flexibility come from? Why do some people seem to have stronger powers of imagination and innovation than others? How does creativity work, and how might we augment it?

Our journey so far has been filled with characters who, through circumstance, have been forced to reach deep and find new reserves of strength and creatively adjust to seemingly insurmountable obstacles, as well as the characters who have helped them do so. So, in this our final chapter, we'll begin with the story of an individual whose own misfortune in the end also unmasked an unexpected store of resilience, and in the process is providing new insights into the human mind, helping to change what leading scientific minds think might be possible.

Derek Amato stood above the shallow end of the swimming pool and called for his buddy in the Jacuzzi to toss him the football. Then he launched himself through the air, headfirst, arms outstretched. He figured he could roll onto one shoulder as he snagged the ball, then slide through the top layers of the water to slow his momentum and cushion his fall. It was a grave miscalculation. The tips of

Amato's fingers brushed the pigskin—then his head slammed into the pool's concrete floor with such bone-jarring force that it felt like an explosion. He pushed to the surface, clapping his hands to his head, convinced that the water streaming down his cheeks was blood gushing from his ears.

At the edge of the pool, Amato collapsed into the arms of his friends, Bill Peterson and Rick Sturm. It was 2006, and the thirty-nine-year-old sales trainer was visiting his hometown of Sioux Falls, South Dakota, from Colorado, where he lived. As his two high school buddies drove Amato to his mother's home, unsure what else to do, he drifted in and out of consciousness, insisting that he was a professional baseball player late for spring training in Phoenix. Amato's mother rushed him to the emergency room, where doctors diagnosed Amato with a severe concussion. They sent him home, instructing his mother to wake him every few hours.

It would be weeks before the full impact of Amato's head trauma became apparent: 35 percent hearing loss in one ear, headaches, memory loss. But the most dramatic consequence appeared just four days after his accident. Amato awoke hazy after near-continuous sleep and headed over to Sturm's house. As the two pals sat chatting in Sturm's makeshift music studio, Amato spotted a cheap electric keyboard.

Without thinking, he rose from his chair and sat in front of it. He had never really played the piano—never had the slightest inclination to. Now his fingers seemed to find the keys by instinct and, to his astonishment, ripple across them. His right hand started low, climbing in lyrical chains of triads, skipping across melodic intervals and arpeggios, landing on the high notes, then starting low again and building back up. His left hand followed close behind, laying down bass, picking out harmony. Amato sped up, slowed down, let pensive tones hang in the air, then resolved them into rich chords as if he had been playing for years. When Amato finally looked up, Sturm's eyes were filled with tears.

Amato played for six hours, leaving Sturm's house early the next morning with an unshakable feeling of wonder. He had fooled around with instruments in high school, even learned a decent rhythm guitar. But nothing like this. Amato searched the Internet for an explanation, typing in words like *gifted* and *head trauma*. The results astonished him. He read about Tony Cicoria, an orthopedic surgeon in upstate New York who was struck by lightning while talking to his mother from a telephone booth. Cicoria then became obsessed with classical piano and taught himself how to play and compose music. After being hit in the head with a baseball at age ten, Orlando Serrell could name the day of the week for any given date. A bad fall at age three left Alonzo Clemons with permanent cognitive impairment, Amato learned, and a talent for sculpting intricate replicas of animals.

Finally Amato found the name of Darold Treffert, a world-recognized expert on savant syndrome—a condition in which individuals who are typically mentally impaired demonstrate remarkable skills. Amato fired off an e-mail; soon he had answers. Treffert, now retired from the University of Wisconsin School of Medicine, diagnosed Amato with "acquired savant syndrome." In the seventy-five or so known cases, ordinary people who suffer brain trauma suddenly develop what can seem like almost-superhuman new abilities: artistic brilliance, mathematical mastery, photographic memory, or simply uncommon creativity. One acquired savant, a high school dropout brutally beaten by muggers, is the only known person in the world able to draw complex geometric patterns called "fractals"; he also claims to have discovered a mistake in pi. A stroke transformed another from a mild-mannered chiropractor into a celebrated visual artist whose work has appeared in publications like *The New Yorker* and in gallery shows, and sells for thousands of dollars.

The neurological causes of acquired savant syndrome are poorly understood. But the Internet has made it easier for people like Amato to connect with researchers who study savants, and improved brain-imaging techniques have enabled those scientists to begin to probe the unique neural mechanisms at work. Some have even begun to design experiments that investigate an intriguing possibility: Creative genius lies in all of us, just waiting to be unleashed.

The word *creativity* itself has been used to describe all kinds of things, but cognitive psychologists and neuropsychologists who research it have gone to great lengths in recent years to pin down a definition that allows them to study it and lay the groundwork for a legitimate body of scientific work that tells us something about the brain. Most generally agree that creativity must have two components. Creativity is, first of all, the juxtaposition of disparate ideas in new ways. But this new combination or insight must also be "useful." That requirement, Harvard Medical School's Alice Flaherty writes in a 2005 paper, "captures the distinction between the creative and the merely eccentric or mentally ill (novelty without utility)."

"Using a lever to move a rock might be judged novel in a Cro-Magnon civilization," she notes, "but not in a modern one."

One popular model of creativity holds that the act of creating itself can be divided into four stages: Preparation, Incubation, Illumination (or insight), and Verification (or evaluation). Each stage presumably relies on a different set of neural substrates and can be studied individually.

"Creativity is not just one thing, it's many things," says Rex Jung, a neuroscientist at the University of New Mexico who studies creativity. "It's not just the aha moment. It's not just sitting around and thinking about nothing in particular and letting your mind wander. The multistage process is really important."

Even so, it's the middle two stages that seem to represent that inspired act of birthing something new that we often think of when we talk about creativity. "Incubation" is the loosening up that follows preparation, the letting go that allows disparate ideas to mingle and merge. And "Illumination" is when these new combinations or insights crystallize and enter our conscious minds. It's *that* moment—when great artists and original thinkers are struck with inspiration seemingly out of nowhere—that is the stuff of legend.

The Greek mathematician Archimedes is said to have been relaxing in the bathtub when he had the insight that one could determine volume of an irregular-shaped object by measuring the amount of water it displaced. The revelation was so sudden that Archimedes "leapt out of the vessel in joy, and, returning home naked, cried out with a loud voice that he had found that of which he was in search, for he continued exclaiming, in Greek, εὕρηκα (I have found it out)," the roman architect Vitruvius would famously write.

Isaac Newton is said to have discovered gravity while sitting under an apple tree. Beethoven took daily walks through a Viennese forest after lunch to spur his creativity, jotting down ideas along the way. And generations of writers have been inspired to drink heavily by the example of Ernest Hemingway and his brethren, hoping they can somehow imbibe a little magic.

The story of Derek Amato and others like him is really a story about these seemingly magical idea-generation phases of creativity, Incubation, and Illumination. How is it, exactly, that a bump on the head or a bolt of lightning can suddenly unleash the muse? Where does this new compulsion to create and the ability to do so on command come from? And what does it mean for the rest of us?

Bruce Miller, who directs the UCSF Memory and Aging Center, treats elderly people stricken with Alzheimer's disease and late-life psychosis. One day in the mid-1990s, the son of a patient surprised Miller when he described a strange new symptom overtaking his father: an obsession with painting. Even stranger, as his father's symptoms worsened, the man said, his father's paintings improved. Miller was dubious until the son sent him some samples. The work, Miller recalls, was "brilliant."

"The use of color was striking. He had an obsession with yellow and purple," Miller says. Soon the patient, a brainy businessman with no previous artistic interests, had lost his grip on social norms: He was verbally repetitive, changed clothes in public parking lots, insulted strangers, and shoplifted. But he was winning awards at local art shows.

By 2000, Miller had identified twelve other patients who also displayed unexpected new talents as their neurological degeneration continued. As dementia laid waste to brain regions associated with language, higher-order processing, restraint, and social norms, their artistic abilities exploded. Some had started stealing tips from restaurant tables; one was arrested for skinny-dipping in a public pool; another abused people on the street. But they produced paintings that won awards. They composed quartets. Some could no longer name the objects that they painted. But their art was striking for its realism and technical skill.

Though these symptoms defied conventional wisdom on brain disease in the elderly—artists afflicted with Alzheimer's typically *lose* artistic ability—Miller realized they were consistent with another population described in the literature: savants. Savants often display an obsessive compulsion to perform their special skill, and they exhibit deficits in social and language behaviors, defects present in dementia patients. Miller wondered if there might be neurological similarities, too. Although the exact mechanisms at work in the brains of savants have never been identified and can vary from case to case, several studies dating back to at least the 1970s have found left-hemispheric damage in autistic savants with prodigious artistic, mathematical, and memory skills.

Miller decided to find out precisely where in the left hemisphere of regular savants—whose skills usually become apparent at a very young age—these defects existed. He read the brain scan of a five-year-old autistic savant able to reproduce intricate scenes from memory on an Etch-A-Sketch. Single-photon emission computed tomography (SPECT) showed abnormal inactivity in the anterior temporal lobes of the left hemisphere—exactly the results Miller found in his dementia patients. The exact functions of the anterior temporal lobe remain a matter of debate, but studies suggest parts of it are critical for our ability to recall, contextualize, and categorize objects, people, words, and facts and the processing of meaning.

In most cases, scientists attribute enhanced brain activity to neuroplasticity. Just like those monkeys whose brains devote more cortical real estate to their fingertips when called upon repeatedly to stop spinning disks, artistic skills grow with practice because of the organ's ability to redistribute cortical resources based on frequency of use. But Miller offered a wholly different hypothesis for the mechanisms at work in his patient. Savant skills, Miller argued, emerge because the areas ravaged by disease—which are associated with logic, verbal communication, comprehension, and perhaps social judgments—have actually been *inhibiting* latent artistic abilities present in those people all along. As parts of the logical brain go dark, some of the circuits keeping the parts of the brain associated with creativity in check disappear. The skills do not emerge as a result of newly acquired brain power; they emerge because for the first time, the areas of the brain associated with the free flow of ideas can operate unchecked.

Miller's theory fits in with the work of other neurologists who are increasingly finding cases in which brain damage has spontaneously, and seemingly counterintuitively, led to positive changes—eliminating stuttering, enhancing memory in monkeys and rats, even restoring lost eyesight in animals. In a healthy brain, the ability of different neural circuits to both excite and inhibit one another plays a critical role in efficient function. But in the brains of dementia patients and some autistic savants, Miller argued, the lack of inhibition in areas associated with creativity led to keen artistic expression and an almost compulsive urge to create.

In the weeks after his accident, Derek Amato's mind raced. And his fingers wanted to move. He found himself tapping out patterns, waking up from naps with his fingers drumming against his legs. He bought a keyboard. Without one he felt anxious, overstimulated; once he was able to sit down and play, relief washed over him, followed by a deep sense of calm. He'd shut himself in, sometimes for as long as two to three days, just him and the piano, exploring his new talent, trying to understand it, letting the music pour out of him.

Amato experienced other symptoms, many of them negative. Black and white squares appeared in his vision, as if a transparent filter had synthesized before his eyes, and moved in a circular pattern. He was plagued with headaches. The first one hit three weeks after his accident, but soon Amato was having as many as five a day. They made his head pound, and light and noise were excruciating. One day, he collapsed in his brother's bathroom. On another, he almost passed out in a Wal-Mart.

Still, Amato's feelings were unambiguous. He was certain he had been given a gift. Amato's mother, like many parents, had always told him he was extraordinary, that he was put on the planet to do great things. Yet a series of uninspiring jobs had followed high school—selling cars, delivering mail, doing public relations. He'd reached for the brass ring, to be sure, but it had always eluded him. He'd auditioned for the television show *American Gladiators* and failed the pull-up test. He'd opened a sports management company, handling marketing and endorsements for mixed-martial-arts fighters; it went bust in 2001. Now he had a new path.

The evidence of some sort of God-given gift, Amato believed, lay not just in the ease he felt when he put his fingers on the keyboard. For him, it also came from the drive he felt, the burning compulsion to play. It came with an urgency he had rarely experienced before. He felt it in his heart: This is what he was meant to do.

Amato began planning a marketing campaign. He wanted to be more than an artist, musician, and performer. He wanted to tell his story and inspire people.

Amato's obsessive drive to work on his new skill is something that Darold Treffert and Bruce Miller both commonly see in the savants they consult with. But it is also a defining characteristic of a number of other neurological conditions.

As it happens, Alice Flaherty, the neurologist at Massachusetts General Hospital working with Liss Murphy, has penned several influential reviews on creativity and neurological defects and wrote a 2004 book, *The Midnight Disease: The Drive to Write, Writer's Block, and the Creative Brain*. Flaherty brings a rare set of credentials and insights to the field: She earned a Harvard M.D. in neurology and an MIT neuroscience Ph.D. But then she did postdoctoral work in the school of hard knocks.

In 1998, just a month after finishing her medical school residency, Flaherty gave birth to premature twin boys. Neither one survived. The trauma, the tragedy, the hormones, they did something to Flaherty. After ten days of crushing sadness, she was suddenly overcome with a rare form of mania, which manifested as an overpowering urge to write. Soon she was covering everything around her with words. She wrote on scraps of toilet paper. She wrote on her pants, and up and down her arms. She churned out torrents of prose, scribbling them in minuscule script on tiny Post-it notes in the middle of the night. She wrote while driving.

It was "an overpowering compulsion," she would later say.

"That's all I was conscious of—I had important ideas that I needed to write down because otherwise I would forget them."

Flaherty realized she was manic, "strung out," verbose," and "excited about everything." And even in this state, Flaherty realized almost immediately, there was nothing normal about what was happening to her. Flaherty had read about her condition in medical school. It was called "hypergraphia," an obsessive drive to write. She quickly made a decision.

"I was looking at all this writing I did, which is mostly like garbage high school girl diary kind of writing, and I thought, 'People with hypergraphia, if they don't get published, it's a disease, it's an illness symptom,'" she says. "'If they get published, well, then it's not a disease, then they're writers.'"

Flaherty began to channel her compulsion into writing that would later result in two published books. Somehow from within the vortex of overpowering emotions, Flaherty retained a calculating, professional perspective. But there was also a deeply human component to her choice. Flaherty wanted—she needed—to make sense of it all. She needed not just to write—but to *connect*. When I spoke with Flaherty, I was struck by her language and how much it reminded me of conversations I had with Derek Amato and Bruce Miller.

"I desperately wanted to communicate with people," she told me. "I really had this primary urge to tell everybody everything that I ever thought, all the time. But I also wanted to see if I could even find anyone who had been through what I was going through. I wanted to make people understand that their brains were affecting how they wrote."

People who don't know her, Flaherty once said, often ask her why she calls what happened to her a "disease"—especially since she also thinks of it as a "miracle."

"In part, it is because of the way my writing sucked me away from everything else," she explains. "Also because of how strange it felt to be suddenly propelled into a creative state by what were probably postpartum biochemical changes. I hate to think that writing—one of the most refined, even transcendent talents—should be so influenced by biology. On the other hand, as a neuroscientist, I realized that if we can get a handle on fluctuations in creativity, we might be able to find ways to enhance it."

Flaherty dug out as many medical case studies and historical examples as she could find detailing the experiences of others who'd experienced similar extreme bursts of creativity—both the drive to create, and the mental flexibility to do so with ease. She discovered the work of Bruce Miller and Darold Treffert and their acquired savants. She also consumed the long literature linking mental illness to great works of art. Through reading the work of the psychologist Kay Redfield Jamison and others, she discovered that writers are *ten times* more likely to be bipolar than the rest of the population—poets *forty times.* She also found another interesting connection: While the genetic component of bipolar and schizophrenia seemed to tear through families, destroying the lives of the sickest, it often seemed to inculcate the mildly affected with uncommon drive and unusual ideas, and help those also blessed with talent to go on to ageless renown.

There are many writers and poets who write and rhyme compulsively but whose work would never be considered genius. And there are any number of inspired geniuses who are not bipolar. Still, Flaherty was struck by how often she found what she was looking for—how often, it seemed, mental illness appeared alongside the greats. William James, the writer, philosopher, and psychologist, was mildly bipolar, she discovered. His brother Henry, the renowned author, suffered from depression. Their illnesses seemed to hit a sweet spot, in which they could still function and produce, unlike their lesser-known, less productive bipolar sibling Robert James. Flaherty herself treated William Styron, whose descent into crippling depression resulted in his 1990 book, *Darkness Visible.*

"Styron took being sick really seriously," Flaherty says. "And he really thought about it a lot and how to sort of honor the importance of illness in your life."

Many neurologists now believe that Fyodor Dostoevsky had a rare form of temporal lobe epilepsy only discovered in the 1970s, called "Geschwind syndrome," one of the symptoms of which is hypergraphia. Such a condition would account for his mood swings, his free-floating feelings of doom and ecstasy, and his overpowering desire to write. The author Lewis Carroll also may have had it, too. He wrote 98,721 letters before he died. To Flaherty, the scene in *Alice's Adventures in Wonderland* where Alice falls down the rabbit hole sounded an awful lot like a seizure aura that Gustave Flaubert, also an epileptic, famously wrote about. Other possible candidates for epilepsy she found in a review of the medical and popular literature included Tennyson, Poe, Swinburne, Molière, Pascal, Petrarch, and Dante.

Like Miller, Flaherty noted that all of the brain conditions she found associated with hypergraphia, and many associated with creative bursts in other media, involved regions in the temporal lobes of the brain, which run beneath the frontal and parietal lobes from roughly behind the eyes, past the ears, and stop just short of the back of the head.

The exact neural cascade that defects in this one part of the brain could set off elsewhere were difficult to precisely characterize because the brain is so complex and remains in many ways such a mystery. As we learned in the last chapter, a disturbance in the temporal lobes can travel. It can be akin to throwing a big rock into the middle of a pond: Shock waves ripple out through all its connections to other parts of the brain—like the emotional limbic system, which Flaherty believes may play a key role in the overpowering need to communicate, or the frontal lobes, a movement and idea generator.

But like Miller, Flaherty suspected that it was the *loss* of function caused by these brain conditions, a *release* of "inhibition" centered in circuits rooted in the temporal lobes of the brain, that "unmasked" creative drive and in some cases latent abilities. Temporal lobe epilepsy, she explains, hyperactivates areas in the temporal lobe of the brain during seizures. But the seizures leave scars behind and in between the seizures the areas that were hyperactivated are left stunned. This lethargic state, Flaherty argues, could release normally powerful inhibitory control exerted by portions of the temporal lobe and the circuits they are connected to over other parts of the brain. It's as if the circuits, having survived a storm of extreme activity, go into hibernation to recover.

In bipolar disorder, meanwhile, Flaherty believes a dysregulation in the temporal areas of the brain also somehow unleashes other areas of the brain, in this case the primitive emotional areas in the limbic system. This dysregulation fuels the wild vacillation between manic overdrive and the crippling malaise that characterizes manic-depressive episodes.

Miller's most artistic dementia patients did not suffer from seizures or bipolar disorder. But they were invariably those whose disease also ravaged portions of their temporal lobes. In their cases it is likely, Flaherty argues, that this release of inhibition on a portion of their relatively intact frontal lobes leads to increased creativity. Though we often think of the frontal lobes as the seat of reason and logic, some parts of the frontal areas of the brain are also important in the initial incubation and illumination phases of creativity.* They are where the ideas pop up in our conscious minds when we

* Certainly, there are differences between those whose epilepsy affects the right part of the brain and those whose epilepsy affects the left. People with left-sided temporal lobe epilepsy—which would release right areas—tend to be hypergraphiac and hyperreligious, but also more rule bound and philosophical. Flaherty speculates that the religious aspect comes from the right. People with right-sided temporal lobe epilepsy would tend to get more hallucinatory, emotional effects. The left side, when released, is more manic and imagistic.

"brainstorm"—areas, you might recall, that Paul Reber has begun to explore in his studies of insight and intuition.

Meanwhile, the deteriorating language-processing areas of the temporal lobe that Miller finds in his patients are involved in judging meaning. Their failure could free up these idea-generation areas of the frontal lobe.

"When those areas are overactive it's actually hard to write because you're judging everything," Flaherty says. "You're saying, is this meaningful or not? And you're too critical of your work."

When these areas are inactive—as in the case of people with temporal lobe dementia or temporal lobe epilepsy—the opposite occurs. The filter goes down. Ideas flood into our consciousness.

Some of the most fascinating findings on the neuroscience of creativity in the realm of music come from otolaryngologist Charles Limb and neuroscientist Allen Braun of the National Institutes of Health, who coauthored a 2008 study of jazz musicians. They also shed light on how creativity might function in healthy individuals, and how experts in a specific musical domain who have mastered the chops needed for their craft can use what they know to create new kinds of masterpieces.

Limb, who plays the saxophone, piano, and bass and writes music, had wanted to know how his idols, people like John Coltrane and Charlie Parker, could get up onstage and improvise for seemingly endless jam sessions. Their creativity never seemed to flag. Limb himself had learned to improvise in middle school, he says, and it was only then that he fell thoroughly in love with music and came to "need it." He specialized in studying the functioning of the ear in part because of this passion.

Limb's passion for his art has absolutely nothing to do with a bipolar condition, nor has he experienced brain damage. Limb is highly functional. When I tracked him down he was preparing to move from a post at Johns Hopkins University to a post at UCSF, taking over the endowed chair previously held by Mike Merzenich. There is just something about being in the creative zone that makes him, like many of us, feel alive.

In 2008, while working in Braun's lab, Limb realized there was an opportunity to use the cutting-edge brain-scanning technologies in the facility to examine where this feeling came from and exactly what was happening in his brain when he improvised. To find out, Limb recruited six highly trained, right-handed pianists from the local jazz scene.

Limb had each musician memorize a melody, then instructed them to lie in a brain-scanning machine with a keyboard resting on their lap and play along with a prerecorded backup track piped in through speakers. Then he moved to the second part of the experiment, dispensing with the memorized melody and instructing the musicians to improvise over a chord progression. Limb wanted to know how the brain patterns differed.

During improvisation, Limb and Braun demonstrated, the brain seems to shut down areas involved in self-monitoring and inhibition. Meanwhile, not surprisingly, it turned up, so to speak, areas associated with the senses, which indicated a "heightened sense of awareness."

To New Mexico's Rex Jung, this validated his belief that even in healthy individuals skilled at creativity, the creative process involves a negotiation, a toggling between two distinct circuits of the brain that can inhibit one another. Creative individuals are skilled at letting loose the areas of the brain needed for divergent thinking and novel associations. But there's a second step, one often missing in savants or those with brain damage: They also need to rein it in and harness a separate, more critical system to edit down the ideas in the "evaluation phase" of creativity.

In healthy individuals, Jung says, the Incubation and Illumination phases that grow out of it appear to be associated with the activation of a neural circuit called the "default mode network" (DMN). It's this network that is blocked when we find our creativity stymied—when we're hit with writer's block, for instance. The DMN network is spread out across a wide range of areas in the brain, and the neural pathways that connect it are among the most closely connected and overlapping.

This circuit is most active when our attentional resources are unfocused. We use the DMN when we let our minds "wander," scanning the outside world loosely. But it's also the network of circuits we use when we engage in self-reflection, or daydream. The network is centered on areas involved in memories, like the hippocampus, and the medial prefrontal cortex, which is involved in mental simulations. It makes sense that we would rely on these areas to "brainstorm" and free-associate. Some of these areas, Jung notes, were active when Limb's jazz pianists were performing solos.

But when we move into the "validation" phase of creativity, the phase when we begin to winnow down and edit our ideas, Jung believes we pull in a separate pathway known as the "cognitive control network" (CCN). The CCN often acts in direct opposition to the DMN. The CCN includes key areas of the prefrontal cortex involved in attentional control and executive functions. It's this area that Chris Berka's sharpshooters flipped on and used to filter out internal noise.

Limb himself thinks Jung's theory is "plausible" but believes it is task dependent. (Writing a book, obviously, requires different kinds of thoughts and brain states than playing jazz.) Certainly, he adds, there are patterns of neural activations associated with inspiration. But in jazz, at least, the ingredients one has to draw upon to make a new mix are undeniably also dependent on previous experience—in other words, the Preparation phase of creativity.

But Amato doesn't play jazz, and he has very little training. Where does his ability come from? Few people have followed the emergence of acquired savants like Amato with more interest than Allan Snyder, a neuroscientist at the University of Sydney. Since 1999, Snyder has focused his research on studying how their brains function. He's also pressed further into speculative territory than many neuroscientists feel comfortable doing: He is attempting to produce, spontaneously, the same outstanding abilities in people with undamaged brains.

In 2012, Snyder published what many considered to be his most substantive work. He and his colleagues gave twenty-eight volunteers a geometric puzzle that has stumped laboratory subjects for more than fifty years. The challenge: Connect nine dots, arrayed in three rows of three, using four straight lines without retracing a line or lifting the pen. None of the subjects could solve the problem. Then Snyder and his colleagues used transcranial direct current stimulation (tDCS) to temporarily immobilize the same area of the brain destroyed in the left anterior temporal lobe by dementia in Miller's acquired savants. They simultaneously stimulated areas in the right anterior temporal lobe, making the neurons that were more active in the dementia patients and seemed to be associated with creativity more likely to fire.

After tDCS, more than 40 percent of the participants in Snyder's experiment solved the puzzle. (None of those in a control group given placebo tDCS identified the solution.)

The experiment, Snyder argues, supports the hypothesis that the abilities observed in acquired savants emerge once brain areas normally held in check have become unfettered. The crucial role of the left temporal lobe, he believes, is to filter what would otherwise be a dizzying flood of sensory stimuli, sorting them into previously learned concepts. These concepts, or what Snyder calls "mindsets," allow humans to see a tree instead of all its individual leaves and to recognize words instead of just the letters. "How could we possibly deal with the world if we had to analyze, to completely fathom, every new snapshot?" he says.

Savants can access raw sensory information, normally off-limits to the conscious mind, because the brain's perceptual region isn't functioning, he believes. To solve the nine-dot puzzle, one must extend the lines beyond the square formed by the dots, which requires casting aside preconceived notions of the parameters. "Our whole brain is geared to making predictions so we can function rapidly in this world," Snyder says. "If something naturally helps you get around the filters of these mindsets, that is pretty powerful."

Snyder thinks Amato's musical skill adds to mounting evidence that untapped human potential lies in everyone, accessible with the right tools. And, unlike Limb, he believes that creative genius may be accessible without traditional training, When the nonmusician hears music, he perceives the big picture, melodies. Amato, Snyder says, has a "literal" experience of music—he hears individual notes. Miller's dementia patients have technical artistic skill because they are drawing what they see: details.

Limb's collaborator Allen Braun finds some evidence for this in a 2013 study that extends this same idea. A team at the University of Pennsylvania lab of Sharon Thompson-Schill used tDCS to target areas of the left prefrontal cortex that Braun's artists so efficiently shut down when they improvise in jazz, poetry, and rap.

These areas, the Penn team wrote, were part of the cognitive control network involved in "sensory filtering." And when these areas were inhibited, subjects were better able to look at a picture of an object and brainstorm unusual uses for it.

"Usually the prefrontal cortex pulls everything together and filters out sensory information from the environment, so if you have less of a filter you have more interesting ideas," Braun says. In the case of jazz improvisation and spontaneous rap or poetry, which

Braun has also studied, "the filtering in our case is suspended for internal goals or motions or internally generated thoughts."

Naama Mayseless, currently a postdoc at Stanford University, has published a series of papers with her former advisor, Simone Shamay-Tsoory of the University of Haifa, that takes these ideas even further. Mayseless and Shamay-Tsoory studied a forty-six-year-old accountant with no previous artistic experience who developed a hemorrhage in the temporoparietal areas of the brain. For a time, the patient had problems communicating and understanding speech. Remarkably, just like some of Miller's patients, a few days after being admitted to the hospital, despite no history of artistic expression whatsoever, the accountant developed a strong desire to draw.

"Suddenly, I grew a deep perspective," he told Mayseless. "I understand how to draw things. I understand how to transfer the three dimensional [objects] on to paper, which I couldn't before."

But then something even more amazing and scientifically fortuitous happened. Over the course of the following eight months, the patient's hemorrhage receded. He began to recover his language abilities. As this happened, his urges to draw and paint and his "understanding" of "how to transfer three-dimensional work to paper" began to fade away. Eventually, he lost them altogether.

"Despite recurring attempts to paint," the patient later reported, "I have felt a striking reduction in my ability to paint since my language abilities have improved."

Inspired by this case, Mayseless and Shamay-Tsoory recruited thirty-seven volunteers. After a subject would lie down in an fMRI scanner, the scientists instructed him or her to evaluate different drawings for originality. When the volunteers critically evaluated the drawings, the fMRI scans revealed that a very particular area of the brain lit up with activity—the very same area damaged in the hemorrhage patient, along with several frontal areas. What's more, the more active these areas were inside the fMRI scanner, the fewer ideas the volunteers themselves were able to come up with when asked to take a creativity test outside the scanner.

In a follow-up paper published in 2015, Mayseless and Shamay-Tsoory used tDCS to see if they could actively enhance or inhibit creativity by turning the neural activation in these same areas up and down. Mayseless gave the volunteers a test called the "alternate uses task," offering people a random object such as a pencil and then asking them to brainstorm as many ideas for alternate uses as they could come up with. Examples of creatively used items included a tire (you can stack it with other tires to make a table), a nail (you can use it as a perch for a bird to stand on), or a shoe (you can throw it at a politician).

What they found is that the left frontal areas seemed to be acting to inhibit more temporoparietal areas.

Mayseless believes it is the release of a key area of the frontal lobe that is the important factor in freeing up creativity. But she believes the inhibition is the result of a circuit spanning several brain regions that include both the temporoparietal areas targeted by Snyder and another section of the frontal lobe.

These experiments seem to confirm the idea that creativity can be released. But for me, most do not directly resolve the question of whether there is actually objective artistic skill involved—though the case study of the forty-six-year-old accountant who could suddenly draw in three dimensions does seem to lend some evidence to Snyder's theory that "mindsets" might actually, in some cases, be limiting abilities.

There is another possibility, suggested by Berit Brogaard, a neuroscientist and philosophy professor and director of Brogaard Lab for Multisensory Research at the University of Miami. When brain cells die, she argues, they release a barrage of neurotransmitters, and this deluge of potent chemicals may actually rewire parts of the brain, opening up new neural pathways into areas previously unavailable.

"Our hypothesis is that we have abilities that we cannot access," Brogaard says. "Because they are not conscious to us, we cannot manipulate them. Some reorganization takes place that makes it possible to consciously access information that was there, lying dormant."

Brogaard published a paper exploring the implications of a battery of tests her lab ran on Jason Padgett, the savant who claims to have discovered a mistake in pi and can draw fractals. It revealed damage in the visual-cortex areas involved in detecting motion and boundaries. Areas of the parietal cortex associated with novel visual images, mathematics, and action planning were abnormally active. In Padgett's case, she says, the areas that have become supercharged are next to those that sustained the damage—placing them in the path of the neurotransmitters likely unleashed by the death of so many brain cells.

In Amato's case, Brogaard says, he learned bar chords on a guitar in high school and even played in a garage band. "Obviously he had some interest in music before, and his brain probably recorded some music unconsciously," she says. "He stored memories of music in his brain, but he didn't access them." Somehow the accident provoked a reorganization of neurons that brought them into his conscious mind, Brogaard speculates. It's a theory she hopes to explore with him in the lab.

On a beautiful Southern California day in October, I accompanied Amato and his agent, Melody Pinkerton, up to the penthouse roof deck of Santa Monica's Hotel Shangri-La. Far below us, the city's famous pier jutted into the ocean and the Pacific Coast Highway hugged the coastline. Pinkerton settled next to Amato on a couch, nodding warmly and blinking at him with a doe-eyed smile as three men with handheld cameras circled. They were gathering footage for the pilot of a reality TV series about women trying to make it in Hollywood. Pinkerton is a former contestant on the VH1 reality show *Frank the Entertainer* and has posed for *Playboy*. If the series is green-lit, Amato will make regular appearances as one of her clients.

"My whole life has changed," Amato told her. "I've slowed down, even though I'm racing and producing at a pace that not many people understand, you know? If Beethoven scored five hundred songs a year back in the day and was considered a pretty brilliant mind, and the doctors tell me I'm scoring twenty-five hundred pieces a year, you can see that I'm a little busy."

Amato seemed comfortable with the cameras, despite the pressure. A spot on a reality show would represent a step forward in his career, but not a huge leap. Over the last decade, Amato has been featured in newspapers and television shows around the world. He was one of eight savants featured on a Discovery Channel special in 2010 called *Ingenious Minds,* and he was on PBS's *NOVA*. He appeared on a talk show hosted by his idol, Jeff Probst, also the host of *Survivor.* Amato has even appeared on NBC's *Today* show.

Musical renown (and a payday) has yet to follow. He released his first album in 2007. In 2008, he played in front of several thousand people in New Orleans with the famed jazz-fusion guitarist Stanley Jordan. He was asked to write the score for an independent Japanese documentary. But while Amato's musical prowess never fails to elicit amazement in the media, reviews of his music are mixed. "Some of the reaction is good, some of it's fair, some of it's not so good," he says. "I wouldn't say any of it's great. What I think's going to be great is working with other musicians now."

Still, as we strolled down Santa Monica Boulevard to a sushi restaurant after the filming, he hardly could have seemed happier. At the table, Amato smiled broadly, gestured manically with meaty forearms tattooed with musical notes, and poked the air with his chopsticks for emphasis.

"There's book stuff, there are appearances, performances, charity organizations," he said. "There are TV people, film people, commercial people, background stuff. Shoot, I know I missed about another half dozen. It's like I'm on a plane doing about 972 miles an hour! I'm enjoying every second of the ride!"*

Later, as I drove through the streets of Los Angeles with Amato, it seemed to me that there was something undeniably American about his efforts to seize on his accident—which struck when he was close to forty, staring into the abyss of middle-age mediocrity—and transform himself from an anonymous sales trainer into a com-

---

* Amato hasn't exactly been coy about his desire for fame, mailing packets of material to reporters, sending Facebook requests to fellow acquired savants, and continuously updating his fan page—behavior that has raised some doubts among experts. Rex Jung grew suspicious of Amato after reading about his history as an ultimate-fighting promoter. "I couldn't be more skeptical," he says. Jung has spent time with Alonzo Clemons, the savant who sculpts animals. He believes acquired savantism is a legitimate condition. But he notes Amato does not display other symptoms one would expect. Many savants, Jung says, exhibit "exquisite" computational or artistic capacities, but "almost always at the expense of other things the brain does." There is no way to definitively prove or disprove Amato's claims, but a number of credible scientists are willing to vouch for his authenticity. Andrew Reeves, a neurologist at the Mayo Clinic, conducted MRI scans of Amato's brain for Ingenious Minds. The tests revealed several white spots, which Reeves acknowledges could have been caused by previous concussions. "We knew going in that it was unlikely to show any sort of signature change," Reeves says. But Amato's description of what he experiences "fits too well with how the brain is wired, in terms of what parts are adjacent to what parts, for him to have concocted it, in my opinion." Reeves believes the black and white squares in Amato's field of vision somehow connect to his motor system, indicating an atypical link between the visual and auditory regions of his brain. Most neuroscientists who study creativity are willing to give Amato the benefit of the doubt. "I'm interested when people are so skeptical," says Mass General's Alice Flaherty. "You can say he's just made a career choice. But it's a kind of weird career choice to make."

mercial product, an inspirational symbol of human possibility for potential fans dreaming of grander things. Treffert, Snyder, and others all spoke enthusiastically about unraveling the phenomenon of acquired savantism, in order to one day enable everyone to explore their hidden talents. The Derek Amatos of the world provide a glimpse of that goal.

I can't deny, however, that at times Amato's efforts seem, if not crass, at least a little gimmicky. While channel surfing one time after my visit, I came across Amato performing on a nationally syndicated talk show with a singer who was completely deaf.

During my visit I just wanted to hear him for myself, in person. Was he really a musical genius? After parking on Sunset Boulevard, a few blocks from the storied rock-and-roll shrines of the Roxy and the Viper Room, Amato and I headed into the Standard hotel and followed a bedraggled hipster with an Australian accent through the lobby to a dimly lit bar. In the center of the room sat a grand piano, its ivory keys gleaming. The chairs had been flipped upside down on the tables, and dishes clinked in a nearby kitchen. The club, closed to customers, was all ours. As Amato sat down, the tension seemed to drain from his shoulders.

He closed his eyes, placed his foot on one of the pedals, and began to play. The music that gushed forth was loungey, full of flowery trills, swelling and sweeping up and down the keys in waves of cascading notes—a sticky, emotional kind of music more appropriate for the romantic climax of a movie like *From Here to Eternity* than a gloomy nightclub down the street from the heart of the Sunset Strip. It seemed strangely out of character for a man whose sartorial choices bring to mind eighties hair-band icon Bret Michaels. Amato didn't strike me as prodigious, the kind of rare savant, like Blind Tom Wiggins, whose skills would be impressive even in someone with years of training.

It seemed to me that Amato somehow found a place inside himself he did not have access to before, a place where the music flowed out of him. But it was probably true that he'd need a few more years studying the jazz standards before he'd be able to hold his own onstage with those pros who climbed into Limb's brain scanning machine.

At that moment in that empty club with Amato, however, it didn't seem to matter. There was expression, melody, and, it seemed to me, undeniable skill. And if this could emerge spontaneously in Amato, who's to say what potential might lie dormant in the rest of us?

The fact that he wasn't as good as Herbie Hancock, Thelonious Monk, or Bill Evans actually came as a relief—which might seem surprising as we near the end of a long exploration of all the ways that science can enhance human beings.

Often, while researching this book, I found myself amazed and inspired by what humanity has managed to achieve. But there are some things that would seem smaller if we were able to reduce them to simple mathematical equations or rules of logic. Says Limb, the musician-scientist: "Scientists are not going to resolve the mystery of creativity in art by a few experiments. It'll never happen."

For that, I'm grateful. Personally, I much prefer that ideas like love, beauty, and, yes, creative genius and artistic expression remain a mystery. These are the things, after all, that make us feel most human.

CONCLUSION

When we arrived in the operating room, the eleven-year-old boy was already there, lying calmly on a gurney in a blue hospital gown and a pair of G.I. Joe pajama bottoms decorated to look like jungle fatigues. His parents stood next to him, as a team of anesthesiologists worked quietly nearby, preparing to render him unconscious.

The parents were putting on a brave show for their boy, smiling as they whispered endearments, and stroking his hair while carefully avoiding the shaved patches that had served as entry points for the brain surgery Dr. Emad Eskandar had performed the week before. But the moment their child's eyelids fluttered shut and he faded off, the faces of the parents collapsed into anguish. The mother blinked away tears and put her head in her hands, while a nurse gently rubbed her back, before leading her and her husband out of the room.

It was a beautiful, sunny spring day outside, the kind of day made for eleven-year-old boys to run, yell, and play in grassy parks and fields, which just served to emphasize the reason why we were all standing inside a brightly lit operating room at Massachusetts General Hospital instead. The boy had a progressive form of dystonia, a movement disorder that causes muscles to contract uncontrollably, leading the body to twist and contort involuntarily into unnatural, often painful positions—wrists and feet curved in on themselves, legs frozen in place, back doubled over.

The kid on the table had tasted the joys of a normal childhood, Eskandar told me. But now it was slipping away. His feet and legs were starting to freeze up; it was as if his muscles were slowly turning to stone. He was already having trouble running. Soon it would get worse. Though the disease is not fatal, dystonia in its childhood form can be painful and severe, spreading to major muscle groups around the body and sometimes confining patients to wheelchairs.

"He just wants to be a kid," Eskandar said. "His parents just want him to have a normal childhood and do what kids do."

Few experiences I had while researching this book had a more profound effect on me than spending an entire day following around Emad Eskandar as he performed surgery on patients. I watched up close as the phenomenon I had been exploring for months played out in real time: There, unfolding in front of me in the floodlit operating rooms of Mass General, that single transformative moment in one person's life when, thanks to technology, everything suddenly begins to change. Prior to my visit, I had worried about how I would react to being in the operating room observing a live procedure—would I become nauseous or faint? Instead I was utterly mesmerized. Here were living, breathing human beings asking and answering questions, smiling—even meeting my gaze, eye to eye from across the room. Then they went to sleep and surrendered themselves to Eskandar, who popped them open right in front of me, like a mechanic might pop the hood of a car, and set to work tinkering with the biological machinery powering all the functions that animate a living, breathing, laughing, thinking, loving, feeling, moving human being. At one point, I stood gazing over Eskandar's shoulder as he dug several inches into the gray matter of one patient's brain—which was literally pulsing in time with her heartbeat—and lesioned out a piece of tissue that was causing intractable seizures.

But this case—the case of that eleven-year-old boy—was different. I have a son. I have a daughter. In my mind, I could easily visualize the boy on the table in front of me running and playing on sun-dappled weekend mornings. And the image of his terrified parents stayed with me long after they had left the room. I was unsettled.

Days earlier, Eskandar had drilled two dime-size holes in the top of the boy's skull and inserted electrodes into his brain, the procedure I had earlier watched him perform on a patient with obsessive-compulsive disorder. Now it was time to implant the controller for the brain-stimulating device down by the boy's stomach, and run the wires up, underneath the skin, to the electrodes in the child's brain. The faint intermittent electrical pulses would disrupt the paralyzing signals driving the boy's condition. He would, if all went according to plan, return to normal.

After the parents left, the nurses stapled pieces of cloth directly to the boy up and down his body, then wrapped the exposed skin above the cloth in plastic to catch the blood, using Sharpies to mark the areas for incisions. The largest X was down below the boy's abdomen on his left side, and it was here that I watched Eskandar make his first small cut with a scalpel.

The new direct brain stimulation machine being developed by Draper Laboratory, Eskandar told me, may soon render this part of the procedure obsolete. Its miniaturized "hub," smaller than a cell phone, will fit against the back of the head and connect to titanium "satellites," mounted in the dime-sized burr holes drilled into five different locations on the top of the skull. These will interface directly with the electrodes implanted in the brain. The hair will grow over it. You won't even know it's there.

But that technology is not ready yet. So instead, Eskandar stood in front of a tray of gleaming metallic tools and selected a long metal rod with a curved tip. Then he inserted the rod through the incision down by the boy's abdomen and pushed it under the skin up toward the chest, carving a tunnel through the sleeping child's fascia, the layer of tissue between skin and muscle, running all the way up to the shoulder. Through this passageway Eskandar planned to thread the wires that will connect to the electrodes in the brain.

"It's a little barbaric," Eskandar said to me as he worked. "But it's better than making an incision from bottom to top."

Here on display was a paradox I came across often in my research, though never had I seen it up close and illustrated with such stark emotional force. The biotechnology revolution is allowing wondrous feats of transformation, but in some respects the new technologies remain remarkably primitive—so primitive that one of the surgeons using them can still refer to part of his technique as "barbaric." So primitive they can make you cringe. Despite our science fiction–like progress, we are still, in many respects, in the very early days. It's as if the first Model Ts are rolling off the assembly line, as if floppy disks or eight-track tapes have just debuted.

Yet there seems little question we have crossed some sort of a Rubicon—that finally the promise of this biomedical revolution is beginning to pay off. That it's no longer hype and speculation. That's it's real now.

Throughout this book, we have met scientists who are reverse-engineering the human body and mind down to its smallest individual parts, examining them at a resolution that would have been impossible before, and then using that knowledge to rebuild or change us. This is transforming lives in profound ways that would have been unimaginable just a few years ago. That boy on the gurney would run again. In the months following Eskandar's operation, the misfiring areas of the brain causing the boy's muscles to painfully contract would slowly release their relentless grip. Gnarled limbs would unfold, and that kid's paralyzing limp would give way to a confident stride. There's every reason to believe that kid is a "normal" child again, that he can play tag with his friends, wrestle, ice-skate, ski, and for all intents and purposes leave behind his disability.

Technology set Hugh Herr free as well. In those early days after his accident, Herr was tormented by dreams in which he was running through the cornfields behind his parents' house in Pennsylvania—only to wake up to the stumps of his legs, and the cruel realization that, in all likelihood, he would never run again. Today Herr rock-climbs in the Italian Dolomites and jogs laps around Walden Pond. He is walking again—*really* walking.

Because of Steve Badylak's discovery with "pixie dust," Yancy Morales was able to keep his leg, and Mercedes Soto still has her feet. Instead of amputating, doctors were able to tap into the regenerative powers of their bodies and prompt a kind of healing that would have been impossible in any previous age.

DBS allowed Liss Murphy, so depressed she was bedridden, to rise up again and rejoin the world—to become a mother, and rediscover happiness. Blind Pat Fletcher could gaze out—and "see" mountains, using her ears.

Lee Sweeney has still not cured muscle-wasting disease, or age-related frailty. But the first gene therapies have now been approved in Western nations. Our knowledge of the human genome and how to manipulate it seems to grow with every passing month.

Of course, there is still a long way to go. I saw it in the operating room with Emad Eskandar, when he turned to me with that metal rod in hand and lamented the brutality of the procedure he was about to perform. And I saw it manifested in many other forms. David Jayne almost died when he volunteered to undergo an experimental procedure that offered the possibility of being able to communicate through experimental electrodes. Phil Kennedy, the neurologist who invented them, ended up in emergency surgery himself after courageously (or foolishly, depending on your perspective) having them implanted in his own skull. We're still not particularly close to finding a cure for Alzheimer's disease, or age-related memory loss.

Even Hugh Herr's legs aren't perfect. Only when scientists learn how to connect them directly to his nervous system and allow sensory feedback to flow in the other direction will he truly have limbs that function with something close to the moment-to-moment precision of the real thing.

And though there have been feats performed in the lab using brain–computer interfaces that have been truly mind-blowing—a monkey moving a wheelchair with his mind, a quadriplegic woman using a bionic arm to drink a glass of milk just by thinking—Gerwin Schalk's mentor, Jonathan Wolpaw, summed up the current limitations of these technologies perhaps most eloquently. They are good—while not yet good enough to rely upon in real-world conditions.

"There's no brain–computer interface that you would now want to use to control a wheelchair on the edge of a cliff or to drive in heavy traffic," Wolpaw said. "And until that happens, their use is going to be very limited."

But there is every reason to believe that eventually it will happen. It's likely only a matter of time. Certainly Hugh Herr thinks so.

Herr is partnering with the regenerative medicine pioneer Robert Langer and neuroscientist Ed Boyden to launch a new collaborative effort they call the "Center for Extreme Bionics." The center will aim to target no less than the half of the world's population that suffers from "some form of physical or neurological disability."

"Today we acknowledge—and even 'accept' serious physical and mental impairments as inherent to the human condition," their promotional materials state. "But must these conditions be accepted as 'normal'? What if, instead, through the invention and deployment of novel technologies, we could control biological processes within the body in order to repair or even eradicate them? What if there were no such thing as human disability?"

Gerwin Schalk's visions are even more ambitious. To him, the effort to decode imagined speech, as well as to decode many of the other signals in the brain, is not the destination itself, but a mere guidepost en route to an even grander end point. Someday in the not-too-distant future, Schalk believes, it will be possible to seamlessly integrate the human mind, all of humanity in fact, with computers, enabling any number of new capabilities. The ability, for instance, to instantly calculate complex equations—like the square root of 369,007,892,622, or to instantly access every single fact available on the Web as instantaneously as the memory of what you had for lunch an hour ago.

"You would essentially become part of the machine," Schalk says. "You don't need to type anymore. You don't need to do anything to communicate except think. You would have access, immediate access, to a completely barrierless access to all the information that's available to Google."

You would, in fact, have access to the entire world, to a "hive" mind shared by everybody else who happens to be hooked in at that moment, plus the computational power of all the computers linked into the Net.

"So now you have a billion people, and they're all hooked up like this, and there's no social media where you type in. The world would know what you are about and who you are, right?" Schalk says. "And so all of a sudden you create this supersociety. This would very clearly completely transform, I mean, not only human capacity but also what it means to be human, what it means to be a society, and, I mean, it would completely change absolutely everything about humanity."

"What we're talking about here," he says, "are essentially the absolutely archaic first steps in that direction."

It's a breathtaking vision, one so fantastic that, after contemplating it alongside some of the other technologies explored in this book, it's easy to feel a little unsettled. It's not hard to conjure up visions of a dystopian future. Certainly Hollywood has provided plenty of material to keep the paranoid, cautious, and technophobic busy worrying. What if somebody hacked this hive mind and took it over? Will Hugh Herr's bionics eventually lead us to create evil RoboCops or terminators bent on crushing the rest of us in their cold, metallic hands? Are we headed for a eugenic society, where a genetically engineered superrace lives in marble palaces while the flawed masses clean their floors and toilets? Might an Elmer Schmeisser "thought helmet" be used by a repressive government to monitor our thoughts and silence dissent? Will Emad Eskandar's explorations in deep brain stimulation one day be used for thought control?

These are legitimate concerns, and there is a place for discussion of what we might do to ensure these things do not occur. But for me, in the final analysis, to focus on the dystopian visions so popular in Hollywood—whether you believe them to be sensationalistic, plausible, or bizarrely interesting—and even the optimistic futures projected by scientists like Schalk would be to miss the point.

To me, fears of government mind reading seemed distant indeed next to the story of David Jayne, paralyzed and no longer able to speak, using technology to clown around with his two little children at dinnertime, and tell them he loves them. Or of a blind Pat Fletcher standing in the desert weeping because she could "see" the mountains again—by using her ears. For me, concerns about the remote possibility of a terminator created in some distant future quickly faded when juxtaposed against images of Hugh Herr out for a brisk jog around Walden Pond on prosthetic runner's legs.

Thus again and again I found myself focusing instead on what to me seemed the most powerful ideas and uses for these technologies, and the stories that stayed with me. They all related to what drew me to bioengineering in the first place—that ability to restore to those who have lost them, the things that allow us to express our humanity and connect with the world and those around us. The capabilities that allow human beings to embrace life. To move—to reach out to the world and act upon it, to explore, to dance with a loved one, to climb a rock wall; to sense—to reach out and touch our children and parents, to tell them we love them, to watch a sunset; to think—to create, to feel, to remember good times, to mourn.

I am skeptical that technology will ever fundamentally transform us. In the end, I come back to the same lesson Hugh Herr learned all the way in the beginning. At first he experienced his realization that he could make legs in all shapes and sizes, a liberation. His devices need not look human. But later he was drawn inexorably back to nature's sublime solutions, solutions refined by evolution over billions of years. He realized that the most efficient and elegant solutions to the problems he sought to solve already existed. He realized that the answer was to reverse-engineer the human body so he could understand how it worked. Only then could he rebuild it.

Maybe someday we will have the choice of being stronger, to take a pill to artificially improve our memory, to communicate telepathically or see in the dark. Perhaps we will join a hive mind. But it will take more than that to change that which, at its core, makes us essentially human—that which makes us happy. I saw no sign that we are at all close right now to changing the fundamental human spirit.

If there's a message in this book, it's an optimistic one. These new technologies need not be contemplated with dread, or seen simply as tools to transform or transcend humanity. Any efforts to augment our abilities for our immediate future are not the main point. The most important story is the one about enhancing not our abilities, but our humanity—our ability to do that which in the final analysis makes life meaningful and worth living.

ACKNOWLEDGMENTS

I would have had nothing to put in this book were it not for the many people who so generously and patiently took the time to explain their research, answer my endless questions, and share with me with their personal stories. I am tremendously grateful to all the people mentioned herein.

This project would never have happened without my agent, Eric Lupfer of WME, who first suggested this idea to me. He guided me through the proposal-writing process and provided feedback and encouragement throughout the book's development.

Best of all, thanks to Eric's efforts, I had the opportunity to work with not just one but two of the smartest, most talented book editors a writer could hope for: Denise Oswald and Hilary Redmon of Ecco. They made this book far better than it otherwise might have been. Hilary recognized the potential and provided invaluable feedback and suggestions early on. When she moved on from Ecco, Denise deftly stepped in and helped take the manuscript to the next level.

I also want to thank Emma Janaskie, for her assistance and patience, and Tom Pitoniak, who did a great job copyediting, as well as Ashley Garland, Miriam Parker, and the rest of the team at Ecco. I was lucky to find Sarah Harrison Smith, who fact-checked portions of the manuscript.

In recent years, I have had the privilege of learning from many wonderful magazine editors. Fred Guterl is my science-writing guru, a generous editor, and a good man, whom I feel lucky to know. He assigned me my first science magazine stories at *Newsweek* and later green-lit my first article on the Hugh Herr project, which started this all. Among those who assigned me some of the other stories I drew upon for inspiration and material: Pam Weintraub, Nicole Dyer, Corey Powell, Kevin Berger, Bridget O'Brian, David Craig, Jenny Bogo, Cliff Ransom, David Rotman, Antonio Regalado, and Brian Bergstein.

While writing this book, I often relied on two of the most patient, wonderful, and intelligent women I know for initial feedback: my lovely wife, Sara Diaz, and my saintly mother, Nancy Kline Piore. It was my mother, herself a skilled writer and published author, who first introduced me to the magic and beauty of the written word. Writing has always been her passion. Yet countless times over the course of the last couple of years, she willingly put aside her own work when I rang her up in her little hilltop studio in Woodstock, New York. And without complaint, she patiently listened and provided instant feedback on whatever chapter I happened to be working on that day. Her generosity was remarkable.

My wife, Sara, didn't even have the option of turning off the phone—her office is down the hall from mine. By chapter three, she might rightly have demanded marriage counseling. But she just smiled (most of the time) and told me whether whatever I was working on sounded good or not.

Throughout this project my father, Michael Piore, provided valuable guidance, feedback, and other assistance. Many others were kind enough to read all or part of this book in its various incarnations. I'd like especially to thank Alexis Gelber for reading the manuscript and providing inestimable encouragement and feedback (not to mention hiring me for my dream job at *Newsweek* magazine all those years ago). Joe Onek, William Mesler, Brad Stone, Joshua Wolf Shenk, Mary Carmichael, Anna Kuchment, Sasha Abramsky, and Nicole Dyer also provided invaluable feedback. Carol Lee Kidd and her staff at CLK Transcription saved me hours of work transcribing my interviews. I'd also like to thank Chris Decherd.

Finally, a big thanks to my sister, Ana, who provided words of encouragement, and to my children, Marcus and Natalia, for all those reenergizing moments of laughter. They remind me every day what's really important.

NOTES

INTRODUCTION

2 *It was a discipline* Kendra Cherry, "What Is Humanistic Psychology?," VeryWell, updated April 26, 2016, https://www.verywell.com/what-is-humanistic-psychology-2795242.

Maslow wrote in 1968 Abraham Maslow, *Toward a Psychology of Being* (New York: Van Nostrand, 1962), 5.

5 *"like a thunderbolt"* John D. Enderle and Joseph D. Bronzino, *Introduction to Biomedical Engineering* (Amsterdam and Boston: Elsevier, 2012), 15.

6 *to the European Parliament* C. Coenen, M. Schuijff, M. Smits, P. Klaassen, L. Hennen, M. Rader, et al., "Human Enhancement Study," project commissioned by the European Parliament, Directorate General for internal policies, Policy Dept. A: Economic and Scientific Policy Science and Technology Options Assessments, 2009, 6, accessed May 28, 2016, https://www.itas.kit.edu/downloads/etag_coua09a.pdf.

7 *"original purpose of medicine"* Francis Fukuyama, *Our Posthuman Future: Consequences of the Biotechnology Revolution* (New York, Picador, 2002), 208.

CHAPTER 1: THE BIONIC MAN WHO BUILDS BIONIC PEOPLE

13 *It had already begun* Alison Osius, *Second Ascent: The Story of Hugh Herr* (Harrisburg, PA: Stackpole Books, 1991), 53.

13 *They'd been planning* Ibid., 48.

But as the two Ibid., 53–54.

14 *Herr had taken aim* Ibid., 27–36.

Then he built Hugh Herr, interview with author, Cambridge, Massachusetts, January 2014.

15 *"Do you want to"* Osius, *Second Ascent,* 55–57.

16 *As the two turned* Herr interview, 2014; Jeff Batzer, telephone interview with author, winter 2016.

18 *Back home in* Adam Piore, "The Bionic Man Who Builds Bionic People," *Discover Magazine,* November 2010, accessed May 25, 2016, http://discovermagazine.com/2010/nov/25-bionic-man-who-builds-people.

Hefting himself up Osius, *Second Ascent,* 118.

19 *"I felt more natural"* Piore, "The Bionic Man."

25 *Gosthnian had been* Osius, *Second Ascent,* 194–96.

27 *Cavagna offered a revolutionary theory* G. A. Cavagna, F. P. Saibene, and R. Margaria, "Mechanical Work in Running," *Journal of Applied Physiology* 19, no. 2 (March 1964): 249–56.

It was just theory R. McN. Alexander and Alexandra Vernon, "The Mechanics of Hopping by Kangaroos (Macropodidae)," *Journal of Zoology* (October 1975): 265–303.

29 *In the end Heglund* Giovanni A. Cavagna, Norman C. Heglund, and Richard Taylor, "Mechanical Work in Terrestrial Locomotion: Two Basic Mechanisms for Minimizing Energy Expenditure," *American Journal of Physiology* 233, no. 5 (November 1977): R243–61.

36 *bounce greater distances* P. G. Weyand et al., "Faster Top Running Speeds Are Achieved with Greater Ground Forces Not More Rapid Leg Movements," *Journal of Applied Physiology* 89, no. 5 (November 2000): 1991–99.

37 *Herr had carefully* D. P. Ferris, M. Louie, and C. T. Farley, "Running in the Real World: Adjusting Leg Stiffness for Different Surfaces," *Proceedings of the Royal Society B: Biological Sciences* 265 (1998): 989–94.

41 *In the 1960s* H. C. Bennet-Clark and E. C. A. Lucey, "The Jump of the Flea: A Study of the Energetics and a Model of the Mechanism," *Journal of Experimental Biology* 47, no. 1 (1967): 59–76.

42 *Herr's lead graduate* Piore, "The Bionic Man."

46 *The first mention* Hugh Herr, "Exoskeletons and Orthoses: Classification, Design Challenges, and Future Directions," *Journal of NeuroEngineering and Rehabilitation* 6, no. 1 (2009): 21.

48 *In May 2014* Erico Guizzo, "Dean Kamen's 'Luke Arm' Prosthesis

Receives FDA Approval," *IEEE Spectrum,* May 13, 2014, accessed May 25, 2016, http://spectrum.ieee.org/automaton/biomedical/bionics/dean-kamen-luke-arm-prosthesis-receives-fda-approval.

49 *In 2014, Herr announced* Luke M. Mooney, Elliott J. Rouse, and Hugh M. Herr, "Autonomous Exoskeleton Reduces Metabolic Cost of Human Walking During Load Carriage," *Journal of NeuroEngineering and Rehabilitation* 11 (2014): 80.

CHAPTER 2: THE BIRTH OF BAMM-BAMM

54 *Holding them in* Jeff Alexander, "Mighty Mite: Rare Condition Gives Boy Incredible Strength," *Muskegon (Michigan) Chronicle,* May 20, 2007, 1A.

"He would, literally" Neil Hoekstra, telephone interview with author, May 2014.

55 *mutation found in another baby* Markus Schuelke, et al., "Myostatin Mutation Associated with Gross Muscle Hypertrophy in a Child," *New England Journal of Medicine* 350 (2004): 2682–88, doi:10.1056/NEJMoa040933.

His mother, who also David Epstein, *The Sports Gene: Inside the Science of Extraordinary Athletic Performance* (New York: Current, 2013), 103.

Her grandfather was Ibid., 101.

59 *Sweeney genetically engineered* L. T. Bish et al., "Long-Term Systemic Myostatin Inhibition via Liver-Targeted Gene Transfer in Golden Retriever Muscular Dystrophy," *Human Gene Therapy* 22, no. 11 (December 2011): 1499–1509, doi:10.1089/hum.2011.102.

63 *Olympic sprinters can* Epstein, *The Sports Gene,* 110; Jane E. Brody, "Muscle Twitches Underlie Athletic Prowess," *New York Times,* August 12, 1980, C1; Judith R. Zierath and John A. Hawley, "Skeletal Muscle Fiber Type: Influence on Contractile and Metabolic Properties," *PLoS Biology* 2, no 10 (October 2004): e348, doi:10.1371/journal.pbio.0020348.

64 *By the 1980s, Louis* Victor McElheny, "The Gene Hunter: Louis Kunkel's 30-Year Quest to Diagnose and Cure Muscular Dystrophy," *Harvard Magazine,* March–April 2011, accessed May 25, 2016, http://harvardmagazine.com/2011/03/the-gene-hunter.

68 *In the late 1960s* Richard J. Roberts, "How Restriction Enzymes Became the Workhorses of Molecular Biology," *PNAS* 102, no. 17 (2005): 5905–08, accessed May 25, 2016, http://www.pnas.org/content/102/17/5905.full%3Fsid%3D0ea62490-1689-49cb-915d-c4f482705ed9.

68 *In 1972 Theodore* T. Friedmann and R. Roblin, "Gene Therapy for Human Genetic Disease?," *Science,* March 3, 1972, 949–55, accessed May 25, 2016, http://www.ncbi.nlm.nih.gov/pubmed/5061866.

69 *gene-editing technology* Heidi Ledford, "CRISPR, the Disruptor," *Nature,* June 4, 2015, 20–24, accessed May 25, 2005, http://www.nature.com/news/crispr-the-disruptor-1.17673#/rise.

The technique is allowing Susan Young, "Genome Surgery," *MIT Technology Review,* March/April 2014, accessed May 25, 2016, http://www.technologyreview.com/review/524451/genome-surgery/.

They can make these Ledford, "CRISPR, the Disruptor."

In 1990, a team "Gene Therapy a Revolution, Timeline," National Institutes of Health, accessed May 25, 2016, https://history.nih.gov/exhibits/genetics/sect4.htm.

led by W. French Anderson R. M. Blaese et al., "T Lymphocyte–Directed Gene Therapy for ADA- SCID: Initial Trial Results after 4 Years," *Science,* October 20, 1995, 475–80.

four-year-old girl Jennifer Kahn, "Molest Conviction Unravels Gene Pioneer's Life," *Wired,* September 25, 2007, accessed May 27, 2016, http://www.wired.com/2007/09/ff-anderson/.

70 *As time went on* James M. Wilson, phone interview with author, April 2014; James M. Wilson, "The History and Promise of Gene Therapy: A Scientist's View from the Front Lines of Basic and Applied Clinical Research," *Genetic Engineering & Biotechnology News,* 31, no. 17 (October 1, 2011), accessed May 25, 2016, http://www.genengnews.com/gen-articles/the-history-and-promise-of-gene-therapy/3836/.

It was one of Sweeney's Wilson, interview with author, April 2014.

fatal levels of bad cholesterol Carl Zimmer, "Gene Therapy Emerges from Disgrace to Be the Next Big Thing, Again," *Wired,* August 13, 2013, accessed May 25, 2016, http://www.wired.com/2013/08/the-fall-and-rise-of-gene-therapy-2/.

71 *Finally, four days later* Wilson, phone interview with author, April 2014; Zimmer, "Gene Therapy Emerges from Disgrace."

75 *produces 600 gigabases* Dr. Bicheng Yang, BGI director of communication and public engagement; Scott C. Edmunds, executive editor of BGI Shenzhen's GigaScience; Rena Lam, BGI Health, interviews with author in Hong Kong and Shenzhen, China, June 2014.

79 *a growing body of research* "Anabolic Steroids May Weaken the Heart," WebMD, http://www.webmd.com/fitness-exercise/20100427/anabolic-steroids-may-weaken-the-heart?page=2.

82 *In the older animals* E. R. Barton-Davis, D. I. Shoturma, A. Mu-

saro, N. Rosenthal, and H. L. Sweeney, "Viral Mediated Expression of Insulin-Like Growth Factor I Blocks the Aging-Related Loss of Skeletal Muscle Function," *PNAS* 95, no. 26 (1998):15603–7.

82 *It read* Ibid.

"a couch potato's dream" Kathleen Fackelmann, "Muscle Strength Restored in Mice Treatment Might Combat Aging," *USA Today,* December 15, 1998, 1D.

83 *When Sweeney trotted out* E. M. Swift and Don Yaeger, "Man or Mouse," *Guardian,* September 2, 2001, accessed May 25, 2016 https://www.theguardian.com/sport/2001/sep/02/medicalscience.highereducation.

"power, endurance" Chris Cooper, *Run, Swim, Throw, Cheat: The Science Behind Drugs in Sport* (Oxford: Oxford University Press, 2012), 29.

In the early 2000s Nicholas Wade, "Gene That Governs Pain Perception Is Found," *New York Times,* December 14, 2007, 32; Epstein, *The Sports Gene,* 261.

85 *In 1961, Hughes invented* Amy Shipley, "Drug Testers Have Designs on New Steroid," *Washington Post,* March 8, 2003, D01; Gordon Hughes, telephone interview with author, fall 2011.

In the early 1990s, Arnold Adam Piore, "Juicers, Trippers, and Crocodiles: The Dangerous World of Underground Chemistry," *Discover Magazine,* March 2012, accessed May 26, 2016, http://discovermagazine.com/2012/mar/08-juicers-trippers-crocodiles-dangerous-underground-chemistry.

90 *Lee and his graduate students* Alexandra C. McPherron, Ann M. Lawler, and Se-Jin Lee, "Regulation of Skeletal Muscle Mass in Mice by a New TGF-ß Superfamily Member," *Nature* 387 (1997): 83–90.

cattle called "Belgian Blue" Alexandra C. McPherron, phone interview with author, April 2015; Epstein, *The Sports Gene.*

91 *published a paper* Markus Schuelke et al., "Gross Muscle Hypertrophy in a Child Associated with a Myostatin (GDF-8) Mutation," *New England Journal of Medicine* 350 (2004): 2682–88.

92 *dogs with other mutations* Antonio Regalado, "First Gene-Edited Dogs Reported in China,"*MIT Technology Review,* October 19, 2015, accessed May 25, 2016, https://www.technologyreview.com/s/542616/first-gene-edited-dogs-reported-in-china/.

CHAPTER 3: THE MAN WITH THE PIXIE DUST

95 *It was so gradual at first* Adam Piore, "How Pig Guts Became the Next Bright Hope for Regenerating Human Limbs," *Discover Magazine,*

July–August 2011, accessed May 25, 2016, http://discovermagazine.com/2011/jul-aug/13-how-pig-guts-became-hope-regenerating-human-limbs.

98 *By the eighteenth century* Andrew Pollack, "Missing Limb? Salamander May Have Answer," *New York Times,* September 24, 2002, accessed May 27, 2016, http://www.nytimes.com/2002/09/24/science/missing-limb-salamander-may-haveanswer.html?pagewanted=all&src=pm.

99 *"swollen monster"* Ann B. Parson, *The Proteus Effect: Stem Cells and Their Promise for Medicine* (Washington, DC: Joseph Henry Press, 2004), 26–27.
teratomas have occasionally appeared Ibid.

100 *"They didn't believe"* Piore, "How Pig Guts Became the Next Bright Hope."

108 *In April 1960* Parson, *The Proteus Effect,* 54–63.
We have 25 trillion Ibid., 48.

109 *Still, as McCulloch made* Ibid., 54–63.

110 *Peering into the microscope* Piore, "How Pig Guts Became the Next Bright Hope."

113 *Studies have shown that* F. Mourkioti et al., "IGF-1, Inflammation and Stem Cells: Interactions During Muscle Regeneration," *Trends in Immunology* 10 (2005): 535–42, http://www.ncbi.nlm.nih.gov/pubmed/16109502.
IGF-1 B. Imberti et al., "Insulin-Like Growth Factor-1 Sustains Stem Cell Mediated Renal Repair," *Journal of the American Society of Nephrology* 18, no. 11 (2007): 2921–28, http://jasn.asnjournals.org/content/18/11/2921.full.
Back in the 1980s Adam Piore, "Hearts & Bones: Gordana Vunjak-Novakovic Is at the Forefront of Regenerative Medicine," *Columbia,* Spring 2012, accessed May 25, 2016, http://magazine.columbia.edu/features/spring-2012/hearts-bones.

119 *The lesson was also* Laura Niklason, interview with author, New Haven, Connecticut, May 2014.

120 *Controlling the environment* Piore, "Hearts & Bones."

124 *ways to regenerate lungs* Vunjak-Novakovic, interview with author, New York City, May 2014.

125 *Badylak's growing list* Piore, "How Pig Guts Became the Next Bright Hope."

CHAPTER 4: THE WOMAN WHO CAN SEE WITH HER EARS

140 *testing at Harvard Medical School* A. Amedi et al., "Shape Conveyed by Visual-to-Auditory Sensory Substitution Activates the Lateral Occipital

Complex," *Nature Neuroscience,* published online May 21, 2007, accessed online May 27, 2016, http://brain.huji.ac.il/publications/Amedi_at_al_vOICe_Nature_Neuroscience07.pdf, doi:10.1038/nn1912.

141 *"We see with the brain"* Sandra Blakeslee, "New Tools to Help Patients Reclaim Damaged Senses," *New York Times,* November 23, 2004, F1.

149 *dingy, windowless room* David H. Hubel, "Evolution of Ideas on the Primary Visual Cortex, 1955–1978: A Biased Historical Account," Nobel lecture presented December 8, 1981, Stockholm, Sweden, accessed May 27, 2016 http://www.nobelprize.org/nobel_prizes/medicine/laureates/1981/wiesel-lecture.pdf.

150 *I first saw* Miguel Nicolelis, *Beyond Boundaries: The New Neuroscience of Connecting Brains with Machines—and How It Will Change Our Lives* (New York: St. Martin's Press, 2011), 7.

151 *actual functional organization* Alyssa A. Botelho, "David H. Hubel, Nobel Prize–Winning Neuroscientist, Dies at 87," *Washington Post,* September 23, 2013, accessed May 27, 2016, https://www.washingtonpost.com/local/obituaries/david-h-hubel-nobel-prize-winning-neuroscientist-dies-at-87/2013/09/23/5a227c2c-7167-11e2-ac36-3d8d9dcaa2c2_story.html.

call it a night Hubel, Nobel lecture.

in magazine advertisements Botelho, "David H. Hubel."

152 *Nobel Prize in 1981* David Ottoson, "Award Ceremony Speech," 1981, Nobelprize.org, accessed May 27, 2016, http://www.nobelprize.org/nobel_prizes/medicine/laureates/1981/presentation-speech.html.

154 *summed up by Carla* Carla Shatz, "The Developing Brain," *Scientific American,* September 1, 1992; Carla Shatz, phone interview with author, March, 2015.

157 *affect his worldview* Michael Merzenich, Ph.D., *Soft-Wired: How the New Science of Brain Plasticity Can Change Your Life* (San Francisco: Parnassus, 2013), 18–19.

skin of the body R. M. Paul et al., "Representation of Slowly and Rapidly Adapting Cutaneous Mechanoreceptors of the Hand in Brodmann's Area 3 and 1 of Macaca mulatta," *Brain Research* 36 (1972): 229–49.

on your thumb Norman Doidge, M.D., *The Brain That Changes Itself: Stories of Personal Triumph from the Frontiers of Brain Science* (New York: Penguin, 2007), 53–55.

159 *"elegant, refined pattern"* Merzenich, *Soft-Wired,* 23–28.

162 *with deep interest* Sharon Begley, *Train Your Mind, Change Your Brain: How a New Science Reveals Our Extraordinary Potential to Transform Ourselves* (New York: Ballantine, 2007), 85–86.

163 *of Vanderbilt University* Ibid., 32–38.

anything had changed M. M. Merzenich et al., "Topographic Reorganization of Somatosensory Cortical Areas 3b and 1 in Adult Monkeys Following Restricted Differentiation," *Neuroscience* 8, no. 1 (January 1983): 33–55.

Instead when Merzenich M. M. Merzenich et al., "Progression of Change Following Median Nerve Section in the Cortical Representation of the Hand in Areas 3b and 1 in Adult Owl and Squirrel Monkeys," *Neuroscience* 10, no. 3 (November 1983): 639–65.

164 *named William Jenkins* William M. Jenkins et al., "Functional Reorganization of Primary Somatosensory Cortex in Adult Owl Monkeys After Behaviorally Controlled Tactile Stimulation," *Journal of Neurophysiology* 63, no. 1 (1990): 82–104.

166 *blur of motion* Begley, *Train Your Mind,* 88.

167 *amazing discoveries* Alvaro-Pascual Leone, "Learning About Seeing from the Blind," University of Wisconsin, Paul Bach-y-Rita Memorial Lecture, April 3, 2009, http://videos.med.wisc.edu/videos/8646.

168 *a case study* R. Hamilton et al., "Alexia for Braille Following Bilateral Occipital Stroke in an Early Blind Woman," *Neuroreport* 11, no. 2 (2000): 237–40.

169 *"the same circuitry"* Leone, "Learning About Seeing from the Blind."

172 *connections with their neighbors* Takao Hensch, telephone interview with author, August 2014. See also Jon Bardin, "Neurodevelopment: Unlocking the Brain," *Nature* 487 (July 5, 2012): 24–26, doi:10.1038/487024a.

175 *feeding volunteers Depakote* Judit Gervain et al., "Valproate Reopens Critical-Period Learning of Absolute Pitch," *Frontiers in System Neuroscience* 7 (2013): 102.

CHAPTER 5: SOLDIERS WITH SPIDEY SENSE

179 *had more of an impact* Gary Klein, *Sources of Power: How People Make Decisions* (Cambridge, MA: MIT Press, 1998), Kindle ed., ch. 4; Daniel Kahneman, *Thinking, Fast and Slow* (New York: Farrar, Straus & Giroux, 2011), 11.

180 *British destroyer HMS* Gloucester Klein, *Sources of Power.*

184 *survivors of a Canadian platoon* Mitch Potter, "This Is Not Peacekeeping," *Toronto Star,* September 30, 2006, A01.

194 *"The brain knows more than"* Amy Kruse, telephone interview with author, October 2014; see also Kahneman, *Thinking, Fast and Slow.*

195 *hardly any mistakes* Richard Barker, retired Boeing project manager, interview with author, July 2014; Dylan Schmorrow, interview with author, Northern Virginia, May 2014; Michael P. Snow, Richard A. Barker, Kenneth R. O'Neill, Brad W. Offer, and Richard E. Edwards, "Augmented Cognition in a Prototype Uninhabited Combat Air Vehicle Operator Console," final report to DARPA (paper summarizing work performed by the Boeing team during Phase 3 of the DARPA program Improving Warfighter Information Intake Under Stress).

196 *In the mid-1990s* Jon Bardin, "From Bench to Bunker: How a 1960s Discovery in Neuroscience Spawned a Military Project," *Chronicle of Higher Education,* July 9, 2012, accessed May 27, 2016, http://chronicle.com/article/From-Bench-to-Bunker-/132743/.

"Humans have great general" Adam Piore, "Brain Boost for Information Overload," *Columbia University Record,* October 15, 2010, accessed May 27, 2016, http://engineering.columbia.edu/brain-boost-information-overload.

198 *all 360 degrees* DARPA press release, "Tag Team Threat-Recognition Technology Incorporates Mind, Machine," Targeted News Service, September 18, 2012; Deepak Khosla, senior scientist in HRL's Information System Sciences Laboratory, telephone interview with author, March 2014.

200 *One of the first recorded* Suzanne Corkin, *Permanent Present Tense: The Unforgettable Life of the Amnesiac Patient, H.M.* (New York: Basic Books, 2013), 153.

202 *the border of the star* Ibid., 154–60.

"Well, this is strange" Ibid., 155.

In 1968, Millner decided Ibid., 187–88.

204 *In one paradigm, researchers* Paul Reber, telephone interview with author, November 2014.

206 *Sure enough, the participants* Phan Luu et al., "Reentrant Processing in Intuitive Perception," *PLoS ONE* 5, no. 3 (2010): e9523, doi:10.1371/journal.pone.0009523; Don Tucker, telephone interview with author, November 2014.

208 *In 1984, researchers at* Corkin, *Permanent Present Tense,* 193.

ubiquitous form of implicit R. L. Buckner et al., "Functional Anatomical Studies of Explicit and Implicit Memory Retrieval Tasks," *Journal of Neuroscience* 15 (January 1995): 12–29.

in college students Endel Tulving and Daniel Schacter, "Priming and Human Memory Systems," *Science,* January 19, 1990, 301–6.

215 *The first was an increase* A. Behneman et al., "Neurotechnology to Accelerate Learning: During Marksmanship Training," *IEEE Pulse* 3, no. 1 (2012): 60–63.

CHAPTER 6: THE TELEPATHY TECHNICIAN

219 *neurologists before David Jayne* David Jayne, "The Unexpected Journey," unpublished memoir.

220 *would later recall* Ibid.

221 *one of David Jayne's neurologists* Phil Kennedy, telephone interview with author, November 2015.

227 *Fetz, pulled off a feat* E. E. Fetz, "Operant Conditioning of Cortical Unit Activity," *Science* 163 (1969): 955–58.

"chasing the magnetic moment" E. E. Fetz, telephone interview with author, March 2015.

228 *Certain neurons in the motor* A. P. Georgopoulos et al., "On the Relations Between the Direction of Two-Dimensional Arm Movements and Cell Discharge in Primate Motor Cortex," *Journal of Neuroscience* 2 (1982): 1527–137; A. P. Georgopoulos, telephone interview with author, March 2015.

231 *In 1996, a pioneering* Sherry Baker, "The Rise of the Cyborgs: Melding Humans and Machines to Help the Paralyzed Walk, the Mute Speak, and the Near-Dead Return to Life," *Discover Magazine,* October 2008, accessed May 27, 2016, http://discovermagazine.com/2008/oct/26-rise-of-the-cyborgs.

234 *In 2004, Kennedy implanted* Ibid.

235 *"This is an elephant"* Ibid.

236 *Schalk met Eric Leuthardt* Adam Piore, "The Army's Bold Plan to Turn Soldiers into Telepaths," *Discover Magazine,* April 2011, accessed May 27, 2016, http://discovermagazine.com/2011/apr/15-armys-bold-plan-turn-soldiers-into-telepaths.

238 *By claiming that he could pry* Ibid.

246 *So Kennedy traveled* Adam Piore, "To Study the Brain, a Doctor Puts Himself Under the Knife: How One of the Inventors of Brain-Computer Interfaces Ended Up Getting One Himself," *MIT Technology Review,* November 9, 2015, https://www.technologyreview.com/s/543246/to-study-the-brain-a-doctor-puts-himself-under-the-knife/.

CHAPTER 7: THE BOY WHO REMEMBERS EVERYTHING

257 *Some suggest dementia was* H. Ebbinghaus, *Memory: A Contribution to Experimental Psychology* (1886), ch. 3, Classics in the History of Psy-

chology, accessed May 27, 2016, http://psychclassics.yorku.ca/Ebbinghaus/memory3.htm.

258 *"I simply had to admit"* A. R. Luria, *The Mind of a Mnemonist* (New York: Basic Book, 1968), 11.

259 *room overlooking the roofs* Eric R. Kandel, *In Search of Memory: The Emergence of the New Science of the Mind* (New York: Norton, 2007), 209.

his actions "heroic" C. W. Domanski, "Mysterious 'Monsieur Leborgne': The Mystery of the Famous Patient in the History of Neuropsychology Is Explained," *Journal of the History of the Neurosciences* 22, no. 1 (2013): 47–52, accessed May 27, 2016, http://www.tandfonline.com/doi/full/10.1080/0964704X.2012.667528#.VBm5ebAtC70.

ensure that previous associations Ebbinghaus, *Memory.*

260 *to form a new memory* R. H. Wozniak, Introduction to Ebbinghaus, *Memory,* http://psychclassics.yorku.ca/Ebbinghaus/wozniak.htm#f1.

longer lists required more practice Kandel, *In Search of Memory,* 210.

Memory is cumulative James L. McGaugh, *Memory & Emotion: The Making of Lasting Memories* (New York: Columbia University Press, 2003), 5–6.

261 *and Alfons Pilzecker suggested* Kandel, *In Search of Memory,* 210–12.

capable of lasting a lifetime McGaugh, *Memory & Emotion,* 36–37.

262 *In the 1930s, two* Ibid.

zapped their brains with electricity C. P. Duncan, "The Retroactive Effect of Electroshock on Learning," *Journal of Comparative and Physiological Psychology* 42 (1949): 32–44.

263 *fascinating paper from 1917* K. S. Lashley, "The Effect of Strychnine and Caffeine upon Rate of Learning," *Psychobiology* 1 (1917): 141–70.

his advisor asked incredulously James McGaugh, interview with author in Irvine, California, December 2014.

264 *stimulant caused them to remember* James L. McGaugh and Lewis Petrinovich, "The Effect of Strychnine Sulphate on Maze-Learning," *American Journal of Psychology* 72, no. 1 (March 1959): 99–102.

fifty years examining why J. L. McGaugh, "Memory Consolidation and the Amygdala: A Systems Perspective," *Trends in Neuroscience* 25, no. 9 (September 2002): 456; McGaugh, *Memory & Emotion.*

265 *mild electric shock* Ryan T. LaLumiere, Thea-Vanessa Buen, and James L. McGaugh, "Post-Training Intra-Basolateral Amygdala Infusions of Norepinephrine Enhance Consolidation of Memory for Contextual Fear Conditioning," *Journal of Neuroscience,* 23, no. 17 (2003): 6754–58.

268 *avoid paying taxes* Kathy Kristof, "U.S. Cracks Down on Rich Tax Evaders," *Tulsa World,* June 22, 2008, E3.

paying estate taxes Jeremy Campbell, "Billionaire Has 'Everlasting'

Brainwave to Beat Death Tax," *Evening Standard,* April 19, 1996, 18; Nick Cohen, "Comment: Without Prejudice: Billionaire's Sting on the Mosquito Coast," *Observer,* August 1, 1999, 29.

268 *Mongolian national team* Liana Aghajanian, "Extreme Memory Tournament: Meet the Mongolian Masters of the Mnemonic," *Guardian,* May 6, 2015, http://www.theguardian.com/world/2015/may/06/mongolia-extreme-memory-tournament.

To win the two-day "Memory Athletes and Researchers Collaborate to Dissect Feats of Memory," Association for Psychological Science, May 6, 2015, accessed May 27, 2016, http://www.psychologicalscience.org/index.php/publications/observer/obsonline/memory-athletes-and-researchers-collaborate-to-dissect-feats-of-memory.html.

in the New York Times Benedict Carey, "Remembering, as an Extreme Sport: The Techniques of 'Memory Athletes' Offer Insight to the Mind," *International New York Times,* May 21, 2014, A7.

270 *"a man with a swollen"* Joshua Foer, *Moonwalking with Einstein* (New York: Penguin, 2011), 33.

276 *In a seminal experiment* Kandel, *In Search of Memory,* 282–83; McGaugh, *Memory & Emotion,* 112–13.

several hours to more than a day Kandel, *In Search of Memory,* 282–83.

278 *at a perceived threat* Robert Langreth, "Viagra for the Brain," *Fortune,* February 4, 2002.

279 *These changes were "structural"* Kandel, *In Search of Memory,* 262–76.

280 *in a highly emotional state* Ibid., 265–66.

282 *In 1993, the team* J. C. Yin et al., "Induction of a Dominant Negative CREB Transgene Specifically Blocks Long-Term Memory in Drosophila," *Cell* 79, no. 1 (1994): 49–58.

A colleague named Alcino Silva R. Bourtchuladze et al., "Deficient Long-Term Memory in Mice with a Targeted Mutation of the cAMP-Responsive Element-Binding Protein," *Cell* 79, no. 1 (1994): 59–68.

285 *bizarre e-mail from* Jill Price, *The Woman Who Can't Forget: The Extraordinary Story of Living with the Most Remarkable Memory Known to Science* (New York: Free Press, 2008), 5.

CHAPTER 8: THE SURGEON CONDUCTING THE SYMPHONY

291 *whole greater than its parts* Nicolelis, *Beyond Boundaries,* 7.

292 *"DBS allows us to go"* Adam Piore, "A Shocking Way to Fix the Brain," *MIT Technology Review,* October 8, 2015, accessed May 27, 2016, https://www.technologyreview.com/s/542176/a-shocking-way-to-fix-the-brain/.

299 *These patients* Ibid.

304 *signal originated in a loop* David L. Pauls, "Obsessive-Compulsive Disorder: An Integrative Genetic and Neurobiological Perspective," *Nature Reviews Neuroscience* 15, no. 6 (2014), http://www.nature.com/nrn/journal/v15/n6/fig_tab/nrn3746_F2.html.

307 *journal* Neuron *in 2005* H. Mayberg et al., "Deep Brain Stimulation for Treatment-Resistant Depression," *Neuron* 45, no. 5 (2005): 651–60.

309 *Sitting in Eskandar's lab* Piore, "A Shocking Way to Fix the Brain."

CHAPTER 9: SUDDEN SAVANTS

314 *Derek Amato stood above* Adam Piore, "When Brain Damage Unlocks the Genius Within," *Popular Science,* March 2013, accessed May 27, 2016, http://www.popsci.com/science/article/2013-02/when-brain-damage-unlocks-genius-within.

317 *That requirement* Alice W. Flaherty, "Frontotemporal and Dopaminergic Control of Idea Generation and Creative Drive," *Journal of Comparative Neurology* 493 (2005): 147–53.

Preparation, Incubation, Illumination Maria Popova, "The Art of Thought: A Pioneering 1926 Model of the Four Stages of Creativity," Brain Pickings, accessed May 27, 2016, https://www.brainpickings.org/2013/08/28/the-art-of-thought-graham-wallas-stages/.

318 *Alzheimer's disease and late-life psychosis* Piore, "When Brain Damage Unlocks the Genius Within."

320 *but studies suggest* M. F. Bonner and A. R. Price, "Where Is the Anterior Temporal Lobe and What Does It Do?," *Journal of Neuroscience,* 33, no. 10 (2013): 4213–15; doi:10.1523/JNEUROSCI.0041-13.2013.

Miller offered B. L. Miller et al., "Emergence of Artistic Talent in Frontotemporal Dementia," *Neurology* 51 (1998): 978–82.

with the work of other neurologists Narinder Kapur, *The Paradoxical Brain* (Cambridge: Cambridge University Press, 2011).

321 *In the weeks after his accident* Piore, "When Brain Damage Unlocks the Genius Within."

322 *something that Darold Treffert* Darold A. Treffert, *Islands of Genius: The Bountiful Mind of the Autistic, Acquired, and Sudden Savant* (London and Philadelphia: Jessica Kingsley, 2010).

2004 book Alice W. Flaherty, *The Midnight Disease: The Drive to Write, Writer's Block, and the Creative Brain* (New York: Mariner, 2004).

She wrote on scraps Alice Flaherty, "The Incurable Disease of Writing: A Neurologist Reflects on the Drives and Frustrations of Literary Creativity," interview, *Harvard Medical Bulletin,* December 2003, 24–30.

323 *"sucked me away from everything else"* Ibid.

324 *William James, the writer* Flaherty, *The Midnight Disease,* 65.

325 *Many neurologists now believe* Ibid., 18–48.

327 *2008 study of jazz musicians* Charles J. Limb and Allen R. Braun, "Neural Substrates of Spontaneous Musical Performance: An fMRI Study of Jazz Improvisation," *PLoS ONE* 3, no. 2 (2008): e1679, doi:10.1371/journal.pone.0001679.

seemingly endless jam sessions Nick Zagorski, "Music on the Mind," *Hopkins Medicine Magazine,* Spring/Summer 2008, accessed May 27, 2016, http://www.hopkinsmedicine.org/hmn/s08/feature4.cfm.

329 *a wide range of areas* R. E. Jung et al., "The Structure of Creative Cognition in the Human Brain," *Frontiers in Human Neuroscience* 7 (2013): 330, doi:10.3389/fnhum.2013.00330.

attentional resources are unfocused Kirsten Weir, "To Be More Creative, Cheer Up: The Way to Tap Your Inner Hemingway Is Not How You Think," *Nautilus,* January 15, 2015, accessed May 27, 2016, http://nautil.us/issue/20/creativity/to-be-more-creative-cheer-up.

330 *In 2012, Snyder* Richard P. Chi and Allan W. Snyder, "Brain Stimulation Enables the Solution of An Inherently Difficult Problem," *Neuroscience Letters* 515, no. 2 (2012): 121–24.

331 *in a 2013 study* E. G. Chrysikou et al., "Noninvasive Transcranial Direct Current Stimulation Over the Left Prefrontal Cortex Facilitates Cognitive Flexibility in Tool Use," *Cognitive Neuroscience* 4, no. 2 (2013): 81–89.

332 *Inspired by this case* N. Mayseless et al., "Unleashing Creativity: The Role of Left Temporoparietal Regions in Evaluating and Inhibiting the Generation of Creative Ideas," *Neuropsychologia* 64 (2014): 157–68.

333 *In a follow-up paper* N. Mayseless and S. G. Shamay-Tsoory, "Enhancing Verbal Creativity: Modulating Creativity by Altering the Balance Between Right and Left Inferior Frontal Gyrus with tDCS," *Neuroscience* 291 (2015): 167–76.

334 *"Our hypothesis is that"* Piore, "When Brain Damage Unlocks the Genius Within."

Brogaard published a paper Berit Brogaard, Simo Vanni, and Juha Silvanto, "Seeing Mathematics: Perceptual Experience and Brain Activity in Acquired Synesthesia," *Neurocase* 2012: 1–10, accessed May 30, 2016, http://www.brogaardlab.com/wp-content/uploads/2013/05/Seeing Mathematics.pdf.

On a beautiful Southern California day Ibid.

INDEX

"absolute pitch," 175
Accelerated Learning Program, 215
ACell, 126, 131
acetylcholine, 7, 61–62
Achilles tendons, 26, 28, 30, 33, 40, 42, 50
acquired savant syndrome, 316–17, 330–31
adenosine, 7
adenosine triphosphate (ATP), 62–63
adrenaline, 265, 269
Advanced Brain Monitoring (ABM), 215
"aerobic respiration," 89*n*
Afghanistan, 183–84, 212–13
"aha" moments, 213–14
Alexander, R. McNeill, 27–28, 29
Alice's Adventures in Wonderland (Carroll), 325
alternate uses task, 333
Alzheimer's disease, 257, 318–20
Amato, Derek, 314–17, 321–22, 331, 334–38
Amedi, Amir, 169–71, 176–77
amnesiacs, 199–200, 208
amygdala, 201, 274, 310
amyotrophic lateral sclerosis (ALS). *See* Lou Gehrig's disease
anabolic steroids, 79–80, 83, 84–87
Anderson, W. French, 69–70
ankle, 37–39, 42
anxiety, and DBS, 309–11
Aplysia, 277–78
Archimedes, 318
Armağan, Eşref, 168–69
Armstrong, Lance, 98–99
Army Research Office, U.S., 188, 238–42
Arnold, Patrick, 85–87
Arnold Schwarzenegger golden retriever, 59, 88, 92
Arnold Schwarzenegger mice, 57
athletes, "in the zone," 217
Au, Samuel, 42
auditory cortex, 140, 191, 241, 243–44, 246
augmented cognition, 194–97, 214–15
Austin, Steve, 44, 48
Avatar (movie), 36*n*
axons, 149–52, 173–74

Bach-y-Rita, Paul, 141, 230
Badylak, Stephen, 99–108, 110–13, 125–26, 127, 343
basal ganglia, 212, 295–97
Batzer, Jeff, 13–14, 15–18, 25
Bay Area Laboratory Co-operative (BALCO), 86–88
Beeman, Mark, 213–14
Beethoven, Ludwig, 318
Belgian Blue cattle, 90–91
Benabid, Alim-Louis, 294–95
Bennet-Clark, Henry, 41*n*
Berka, Chris, 215–17, 329
BGI (Beijing Genomics Institute), 72–77
Bieber, Justin, 58
bioengineering, 5–8
"biomimetic," 118
bioreactors, 117–18, 122
bipolar disorder, 324–25, 326
Bliss, Tim, 276
Bolt, Usain, 36–37*n*
Bonds, Barry, 87
bone marrow transplants, 108–9
bones, 23, 24, 113
Boston Red Sox, 36
Boston University, 235
Boyden, Ed, 344
braille, 165–69
brain-computer interfaces (BCI), 226–27, 228–29, 237–38
BRAIN Initiative, 300
brain myth, ten percent of the, 192–93
brain plasticity. *See* neuroplasticity
Braun, Allen, 327–28, 331–32
Brogaard, Berit, 333–34
Brown, Jim, 2
Buchanan, James, 258
Burke, John, 93
calf muscles, 30, 38, 40, 42
Cambodia, 2–4
Cameron, James, 36*n*
Camp Pendleton, 215–16
Can't Remember What I Forgot (Halpern), 285
cardiac cells, 120–22
cardiomyoplasty, 102
Carrigan, Kim, 14
Carroll, Lewis, 325
Cas9 enzyme, 69
catapult, the, 26
Cathy (patient), 223–25, 236–37, 245
Cavagna, Giovanni, 27, 29, 30
Cecere, Sue Ann, 219, 248
Center for Extreme Bionics, 344
cerebral cortex, 150–51
Chambers, Dwain, 87
Chang, Chris, 73–77
Chappelle, David, 187
Chattahoochee River, 220
children, brain plasticity in, 171–73
chondroitin sulfate proteoglycans (CSPGs), 173–74
Cicoria, Tony, 316
Claparède, Édouard, 200–201
Clemons, Alonzo, 316, 336*n*
Cline, Hollis, 153–54
"cochlea," 159
cochlear implants, 156–57, 158–61, 191
cognitive control network (CCN), 329
Cognitive Technology Threat Warning System, 198–99
Cohn, Joseph, 184–86, 191–93, 199–200, 205–8, 214
Cold Spring Harbor Laboratory, 267, 281, 283–84
collagen, 105–6, 107

Coltrane, John, 327
consciousness, 192–93
consciousness studies, 142–43
consolidation hypothesis, 261–62
Conte, Victor, 86–87
Cooper, Bradley, 192
Cooper, Chris, 83
Corkin, Suzanne, 201, 203
cortically coupled computer vision (C3Vision), 197–98
creativity, 313–38; Amato's case, 314–17, 321–22, 331, 334–38; definition of, 317; Flaherty's research on, 305–6, 322–27; Limb's research on, 327–29; Mayseless and Shamay-Tsoory experiment, 332–33; Miller's theory of, 318–21, 324–27; stages of, 317–18
CREB, 279–84
Crick, Francis, 267
CRISPR, 69, 71, 77, 89, 92
cryptic peptides, 107–8, 112
cyclic AMP, 279

Daedalus, 7
Darkness Visible (Styron), 324–25
Dart, Kenneth, 267–68, 277
Dart NeuroScience, 255–56, 266–72, 285
da Vinci, Leonardo, 24
daydreams, 329
declarative (explicit) memory, 200–201, 203, 275
deep brain stimulation (DBS), 290–311, 341; Dougherty's research, 290–91, 292, 300–301, 305, 308–11; Eskandar's research and surgery, 290–91, 292, 294, 298–305, 308–11, 339–42; Murphy's case, 289–91, 305–7
"Deep Breath," 124–25
default mode network (DMN), 329
Defense Advanced Research Projects Agency (DARPA), 48, 185–86, 194, 197, 198
DEKA Research & Development Corporation, 48–49
Delgado, Jose Manuel Rodriguez, 308–9*n*
DeLong, Mahlon, 295–97
dementia, 257, 319–21, 326, 327, 330, 331
dendrites, 149, 174
Depakote, 175
depression, 289–90, 289–91; Murphy's case, 289–91, 305–7
depth perception, 139, 140
"desynchronizing," 297–98
DNA, 58, 66, 68–69, 267
DNA sequencers, 72, 73, 75–76
Dostoevsky, Fyodor, 325
Dougherty, Darin, 290–91, 292, 300–301, 305, 308–11
Draper Laboratory, 301–2
Duchenne muscular dystrophy, 64–67, 78–79
dyslexia, 156
dystonia, 339–40
dystrophin, 64–67, 78–79
D'Zmura, Mike, 242

EA Sports, 36
Ebbinghaus, Hermann, 259–61
Ebert, Roger, 130
elastic recoil energy in legs, 27, 44
electricity in the body, 39, 61, 128
electroconvulsive therapy (ECT), 262, 263, 279, 290–91, 293
electrocorticography (ECoG), 236, 238, 241, 242, 243

electroencephalography (EEG), 196–97, 242
electromyography (EMG), 43, 247–48
emotion and memory, 264–66
endurance, 83, 88–89*n*
epilepsy, 201, 223, 236–37, 240–41, 294
episodic memory, 273–79, 275*n*
Epstein, David, 84–85, 89
Ervin, Frank, 308–9*n*
erythropoietin (EPO), 83, 89*n*
Eskandar, Emad, 290–91, 292, 294, 298–305, 308–11, 339–42
ethical issues, 7–8, 57–59, 73–74, 87–88, 345–47
exoskeleton, 45–47, 49–52
expertise and intuition, 190–91
extracellular matrix (ECM), 97, 104–8, 111–13, 125–27, 132–33
Extreme Memory Tournament (XMT), 268–69, 270–71
"eye music," 176

Farley, Claire, 37
Ferrell, Robert, 55, 92
Ferris, Dan, 37
Fetz, Eberhard, 227–29
fibronectin, 105–6
fight-or-flight response, 79, 205, 207, 265
filaments, 61–62
firefighters, intuition in, 179–80, 188–90, 191, 204–5
Flaherty, Alice, 305–6, 317, 322–27
"flashbulb memories," 280
Flaubert, Gustave, 325
flea locomotion, 41, 41*n*
Fletcher, Pat, 137–40, 142–44, 161–62, 169–71, 175–76, 343
Flintstones, The (cartoon), 54
Foer, Joshua, 269*n*
Food and Drug Administration (FDA), 48, 106, 235; DBS and, 298, 303, 308, 311
forgetting curve, 260
fractals, 316, 334
Freed, Lisa, 116–18
Freud, Sigmund, 2, 193
Fried, Itzhak, 293–94*n*
Friedmann, Theodore, 68–69, 71
fruit flies, 272–73, 281–83
Fukuyama, Francis, 7
functional near-infrared spectroscopy, 196–97

GDF-8 gene, 55
Geddes, Leslie, 101
Gelsinger, Jesse, 70–71, 92, 124
Genabol, 85–86
gene doping, 84–89
generalized anxiety disorder (GAD), and DBS, 309–11
gene therapies, 58–59, 68–72, 92–94
genetic engineering, 53–94; BGI's research on, 72–77; Hoekstra's gene mutation, 54–56, 91–92; Lee's research on, 84–85, 89–91; Sweeney's research on, 56–67, 77–85, 92
genome, 71–72
Georgopoulos, Apostolos, 228–29, 231
"Geschwind syndrome," 325
Giambi, Jason, 87
Global Positioning System (GPS), 185
Gloucester, HMS, 180–81, 190
Golgi tendon organ, 51
Gosthnian, Barry, 25–26
growth factors, 106

guide RNA, 69
gut feelings, 205–5, 213

Halpern, Sue, 285
Happy (dog), 27–28
Hausler, Jake, 258–59, 287–88
hearing aids, 159. *See also* cochlear implants
heart rate deceleration, 215–16
heart tissue, 99, 120–22
Hebb, Donald, 154, 202, 275
Hebbian learning, 154, 161, 202, 210, 275, 276
heel spring, 32
Heglund, Norm, 29–30
Helicon Therapeutics, 267, 277, 284–85
Hemingway, Ernest, 318
Hensch, Takao, 172–73, 174–76
Hernandez, Isaias, 95–97, 99, 122
Herr, Hugh, 4, 13–23; biomechanics and bioengineering of, 25–26, 31–34, 37–41, 43–44, 45–46, 347; bionic prostheses of, 21–22, 25–26, 34–43, 46, 51, 53, 230, 343–44; initial prosthetic creations, 18–21, 24–25; Mount Washington accident, 13–14, 15–18; potential for human augmentation, 45–47, 49–52, 344
hierarchy of needs, 2, 4
highly superior autobiographical memory (HSAM), 286–87
hippocampus, 200, 201–2, 274, 276, 329
Hoekstra, Liam, 54–56, 91–92
Horgan, John, 308–9*n*
horse locomotion, 27–28
HRL Contracting, 198
Hubel, David, 148–56, 171, 174, 228
Hughes, Gordon, 85–87
human growth hormone (HGH), 79, 80
hypergraphia, 323–24, 325

IGF-1 (insulin-like growth factor 1), 80–83, 89, 90, 92, 113
illumination stage of creativity, 318, 329
imagined speech, 239–43, 246–40, 344–45
implicit learning, 200–211
incubation stage of creativity, 317–18, 329
Ingenious Minds (documentary), 335
inhibition and creativity, 320–21, 325–27
In Search of Memory (Kandel), 278–79
insight, 213–14
intelligence (IQ), 73–74
International Space Station (ISS), 117–18
intestine, 102–6
intracranial EEG (iEEG), 236
intuition, 179–217; Cohn's research on, 184–86, 191–93, 199–200, 205–8, 214; implicit learning and, 200–211; Klein's research on, 180, 181–82, 186–92, 204–5; Reber's research on, 209–14; in soldiers and firefighters, 179–91, 204–5, 211–13, 215–17
"ion channels," 128
Iraq war, 183–84
Iron Cross, 54
Iron Man, 45
Itamura, John, 106–7, 110

James Bond (character), 176–77
James, Henry, 324

James, LeBron, 36
James, Robert, 324
James, William, 259, 324
Jayne, David, 219–21, 232–35, 247–48, 343
Jayne, Melissa, 219–20, 221, 232, 233
Jenkins, William, 164–65
Jeopardy! (TV show), 271
Johansson, Scarlett, 192
Jones, Marion, 87
Jordan, Stanley, 335
Jung, Rex, 317, 328–29, 336*n*

Kaas, Jon, 163–64
Kamen, Dean, 48–49
Kandel, Eric, 277–81, 284–85
Kaplan, David, 127–28
Kennedy, Phil, 231–32, 233–35, 246–47, 248
Khmer Rouge, 2–3
Klein, Gary, 180, 181–82, 186–92, 204–5
knee angle, 35
Knight, Robert, 245
Kobayashi, Hiroshi, 46–47
Kruse, Amy, 194–95, 196, 198, 215
Kunkel, Louis M., 64

laminin, 105–6
Langer, Robert, 113–16, 118, 344
Langreth, Robert, 277–78
learning curve, 260–61
learning pills, 171, 175
Lee, Se-Jin, 84–85, 89–91
Leuthardt, Eric, 236–37, 240–41
Levin, Michael, 128–29
Lewis, Carl, 36*n*
ligaments, 23, 24
Limb, Charles, 327–29
limbic system, 205–6
Limitless (movie), 192
lobsters, 98
locomotion, 27–30, 32–33, 35–36
"logistic template," 129–30
Lomo, Terje, 276
Londono, Ricardo, 112
long-term memory, 278–79
long-term potentiation (LTP), 276
Lou Gehrig's disease, 221–22, 230–32; Jayne's case, 219–21, 232–35, 247–48
Lucy (movie), 192
lungs, 123–25
Luria, Alexander, 258
Luu, Phan, 205–7, 205*n*

Madoff, Bernard, 77
Maffei, Lamberto, 173–74
magnetoencephalography (MEG), 242–43
Mallow, Johannes, 268
Malone, Frank, 19
Manson, Charles, 77
Mantyranta, Eero, 88–89*n*
Maslow, Abraham, 2–3, 4, 127
Mayberg, Helen, 307–8
Mayseless, Naama, 332–33
McCulloch, Ernest, 108–11
McGaugh, James, 262–66, 285–88
McGwire, Mark, 86
McMahon, Thomas, 32–33, 37
McPherron, Alexandra, 90*n*
Meijer, Peter, 142–48
memorization pills, 256, 267–68, 276–77, 292
memory, 253–88; biochemical and genetic basis of, 267–69, 272–77; Ebbinghaus's research on, 259–61; Hausler's case, 258–59, 287–88;

Kandel and, 277–81, 284–85; McGaugh's research on, 262–66, 285–88; repetition and, 260–62; Tully's case, 253–56, 266–73, 276–77, 281–85
memory consolidation, 261–66, 278–80
memory loss, 256–58, 261–62
Memory Pharmaceuticals, 277
Merzenich, Michael, 157–61, 162–65, 191, 296
"method of loci," 269–70, 269*n*
Michelson, Robin, 158–59
Midnight Disease, The (Flaherty), 322
Miller, Bruce, 318–21, 324–27
Milner, Brenda, 201–3, 206
"mindsets," 330–31, 333
mitochondria, 89*n*
Molaison, Henry "H.M.", 201–3, 274–75
molecular brakes, 172, 174–75
Mongols, 26
monoamine oxidase inhibitors (MAOIs), 290
Moonwalking with Einstein (Foer), 269*n*
Morales, Yancy, 131–34, 343
motion capture systems, 35–37, 36*n*
motor cortex, 166–67, 228–31
Mount Washington, 13–14, 15–18
Müller, Georg, 261
Murphy, Liss, 289–91, 305–7, 343
Murphy, Scott, 306
muscles, 23, 60–63; function during locomotion, 28–29, 30–31; genetics and, 60–66
"muscle suit," 8, 46–47
muscle wasting, 57–58, 64–67, 77–84, 92
muscular dystrophy, 64–67, 78–79
myosin, 61–62, 64, 66–67
myostatin, 55, 89–91, 100, 113

National Aeronautics and Space Administration (NASA), 117–18
National Institutes of Health (NIH), 69–70, 101, 162–63, 272, 303, 327
National Photographic Interpretation Center (NPIC), 196
"naturalistic decision making," 181
neuromodulators, 172–73
"neuron doctrine," 274*n*
neurons, 149–56, 172–73, 210–11, 222, 228–31, 274–75, 291–92
neuroplasticity, 137–77, 320; Fletcher's experience, 137–40, 142–44, 161–62, 169–71; Hubel and Wiesel experiments, 148–56, 171, 174; Meijer's research on, 142–48; Merzenich's research on, 157–65; Pascual-Leone's research on, 147–48, 161–62, 165–71
Newton, Isaac, 318
Nicolelis, Miguel, 150
Nike, 32–33
Niklason, Laura, 119–20, 122, 123–25
Nobel Prize, 149, 152, 164, 277
Nogo receptor, 175
"nonsense" words and memory, 259–60
norbolethone, 85–86

Obama, Barack, 300
obsessive-compulsive disorder (OCD), 286–87, 291, 299–300, 303, 304–5
orbital frontal cortex, 304
Osius, Alison, 19
Ottoson, David, 152
Our Posthuman Future (Fukuyama), 7

P300 response, 197–99
pacemakers, 121–22
Padgett, Jason, 334
pain tolerance, 83–84
parietal cortex, 215, 293, 294
Parker, Charlie, 327
Parkinson's disease, 291, 297–98, 304
Pascual-Leone, Alvaro, 147–48, 161–62, 165–71, 191, 293
pattern matching, 181, 189–92
Peek, Kim, 258
Penfield, Wilder, 202
Pentagon, 182–86, 194–95
perineuronal nets (PNN), 174–75
peripheral nerves, 61, 157, 173, 294
Permanent Present Tense (Corkin), 201
Persian Gulf War, 180–81
Peterson, Bill, 315
Philips, 146
pig bladder, 96–97, 100
pig intestine, 104–5
Pilzecker, Alfons, 261
Pinkerton, Melody, 334–35
Pink Floyd, 245
Pistorius, Oscar, 44–45
"pixie dust," 96–97, 126–27
placebo effects, 308
Poeppel, David, 242–44
positron emission tomography (PET), 166–67
posttraumatic stress disorder (PTSD), and DBS, 309–11
Potter, Mary "Molly," 193–94, 194*n*
preparation stage of creativity, 317, 329
Price, Barbara, 57–58
Price, Jill, 286, 288
"primacy" effect, 260
"priming," 208–9
Probst, Jeff, 335
protein kinase A, 279
Proust, Marcel, 275*n*
"pruning," 155
Publius Flavius Vegetius Renatus, 26
Pyc, Mary, 270–71

Radisic, Milica, 121
Ramón y Cajal, Santiago, 274*n*
Ramsey, Erik, 234–35
rapid serial visual presentation (RSVP), 193–94
Ray, Johnny, 231–32
Reber, Paul, 209–14
"recency" effect, 260
recombinant DNA, 68–69
red blood cells, 108–10
Reeve, Christopher, 173, 174
Reeves, Andrew, 336*n*
regenerative medicine, 95–134; Badylak's research on, 99–108, 110–13, 125–26, 127; Hernandez's case, 95–97, 99, 122; Niklason's research on, 119–20, 122, 123–25; Vunjak-Novakovic's research on, 113–22, 124–25, 129–30
Reing, Janet, 111–12
Reinhard, Simon, 268
Remembrance of Things Past (Proust), 275*n*
repetition and memory, 260–62
"repetition suppression," 209, 212
Restoring Active Memory (RAM), 293–94*n*
Richardson, Mattie Theo, 77–78, 79, 82
Riley, Michael, 180–81, 190
Ritaccio, Anthony, 237
Roblin, Richard, 68–69, 71
Rocky (dog), 101, 102–4
Rodriguez, Eugenio, 130–33

Roediger, Henry "Roddy," III, 271–72, 287–88
Romanowski, Bill, 87
Rubinstein-Taybi syndrome (RTS), 284
Run, Swim, Throw, Cheat (Cooper), 83
running, 29–30, 32–33, 36–37, 44–45
running shoes, 32

Sadato, Norihiro, 166–67
Sajda, Paul, 196–98
salamanders, 98, 125, 129
savant syndrome, 316–38, 330–31
Schalk, Gerwin, 223–27, 236–41, 244–46, 344–45
schizophrenia, 300, 311, 324
Schmeisser, Elmar, 238–42, 244
Schmorrow, Dylan, 195–97
Schuelke, Markus, 91
Scoville, William, 201
selective serotonin reuptake inhibitors (SSRIs), 290
"self-actualization," 2–3
senses, 140–42
sensory substitution, 142–43
September 11 attacks (9/11), 142, 262, 280
Serkis, Andy, 36*n*
serotonin, 278–79, 290
Serrell, Orlando, 316
Shaible, Adam, 161–62, 170–71
Shamay-Tsoory, Simone, 332–33
Shatz, Carla, 154
Shereshevsky, Solomon Veniaminovich, 258, 259, 269–70, 280–81
"short-term conceptual memory," 193
single-photon emission computed tomography (SPECT), 320
single-unit recordings, 149–50, 166, 228, 238
Six Million Dollar Man (TV show), 44, 48
"sixth sense," 180, 182–83
Skylark of Space, The (Smith), 239
Smith, E. E. "Doc," 239
Snyder, Allan, 330–31
sodium channel N9A (SCN9A), 84
somatosensory cortex, 157–58, 163–65, 166
Soto, Mercedes, 132–33, 343
soundscapes, 138–39, 140
Sources of Power (Klein), 180
spacing effect, 260–61
spatial memory, 195, 196
spectrograms, 144–46
Spider-Man suits, 31, 44
"Spidey sense," 181, 182–86, 190
Spievack, Alan, 125–27, 131
Spievack, Lee, 96, 126–27
Sports Gene, The (Epstein), 84–85, 89
sprinters, 36–37, 36–37*n*, 44–45
Squire, Peter, 184, 211–13, 222
Starr, Philip, 297–98
stem cells, 5, 80, 110–15, 123–24, 129–30
stress hormones, 265, 280
strychnine, 263–64
Sturm, Rick, 315–16
Styron, William, 324–25
subgenual cingulate, 307
subthalamic nucleus, 297
"Super Baby," 89, 91
Susquehanna River, 19, 20
Sweeney, H. Lee, 56–67, 343; background of, 59–60; ethical concerns of, 58–59, 87–88; gene therapy research, 60–66, 77–84

synapses, 149–56, 210–11, 279
synesthesia, 270

talk therapy, 290
Taylor, C. Richard, 29
temporal lobe epilepsy, 201–2, 325–27, 326*n*
tendons, 23, 26–31
ten percent of the brain myth, 192–93
teratomas, 99, 110
testosterone, 7, 79–80, 86
thalamus, 294, 296, 304
"theta band," 207, 215–16
Thomas, Tammy, 86
Thompson-Schill, Sharon, 331
Tian, Xing, 242–43
Till, James, 109–10
tissue engineering, 113–22
"tongue camera," 142
torque, 39, 50
transcranial direct current stimulation (TDCS), 293–94, 330–33
transforming growth factor beta superfamily (TGF-beta), 90*n*
Treffert, Darold, 316, 324, 337
tricyclics, 290
Tucker, Don, 205–7
Tully, Tim, 253–56, 266–73, 276–77, 281–85, 287
Tumur, Enkhjin, 268
Turing, Alan, 223
two-photon microscopy, 153–54
type I ("slow-twitch") muscle fibers, 62–63, 89*n*
type II ("fast-twitch") muscle fibers, 62–63

unconscious, 2, 192–93
unconscious perception, 193–94

van der Smagt, Patrick, 24, 47–48
Vicon, 35–36
Vietnam War, 183
visual cortex, 148–49, 151, 167, 168, 174–75, 209, 210, 212, 228, 246
Vitruvian Man, 24
Vitruvius, 318
vOICe, 138, 142–46, 170–73; Fletcher's case, 142–44, 161–62, 170–71, 175–76; James Bond (character), 176–77
voice restoration, 221–25, 240–44; Jayne's case, 233–35, 247–48
Voltaire, 98
Vunjak-Novakovic, Gordana, 113–22, 124–25, 129–30

walking, 29–30, 32–33
Watson, James, 267, 277, 283–84
Weyand, Peter, 36–37, 44–45
Wiesel, Torsten, 148–56, 171, 174, 228
Wiggins, Thomas "Blind Tom," 258
Wilson, James, 70–71, 78–79, 124
Wolf, Steven, 96–97
Wolpaw, Jonathan, 226–27, 237–38, 344
Woods, C. Geoffrey, 84
working memory, 195–96, 260, 270
World Anti-Doping Agency (WADA), 58–59, 87–88
Wyeth Pharmaceuticals, 85

Yagn, Nicholas, 46
Yin, Jerry, 281–83